蔡潔儀

百味料理

Oil Cooking

煎炒炸

蔡潔儀 編著　萬里機構・飲食天地出版社出版

蔡潔儀百味料理：煎炒炸

編著
蔡潔儀

編輯
郭麗眉

攝影
幸浩生

翻譯
子惠

封面設計
任霜兒

版面設計
何秋雲

出版
萬里機構・飲食天地出版社
香港鰂魚涌英皇道1065號東達中心1305室
電話：2564 7511　　傳真：2565 5539
網址：http://www.wanlibk.com

發行
香港聯合書刊物流有限公司
香港新界大埔汀麗路36號中華商務印刷大廈3字樓
電話：2150 2100　　傳真：2407 3062
電郵：info@suplogistics.com.hk

承印
凸版印刷（香港）有限公司

出版日期
二〇一一年二月第一次印刷

前言

　　為了提高學習者的烹飪技術水平，培養烹飪技藝的新秀，我根據多年教烹飪的經驗，編寫了這本以技法為主題的食譜，突出了教材的實用性和普及性。為了讓讀者容易掌握，我採用了實踐的方法講述一些菜餚的知識。

　　中國的烹調技法多種多樣，蒸、煮、煎、炒、炸中，水烹及氣烹的歷史最悠久，所以我也從這兩種技法開始，出版了《蒸炆煮燉》；這次我要介紹應用最廣泛的油烹法，編成《煎炒炸》一書。書裏我選了50個食譜，注入了烹調的新元素，加以說明。希望各位讀者能一邊閱讀和一邊應用，把烹調技巧提升到更高境界。

蔡潔儀

目錄

甘脆濃腴的油烹法

炸

油浸

烹

熘

中式烹調基本流程概覽

中國飲食文化源遠流長，粗略估算已有五千多年歷史，聞名遐邇；尤以烹調技法縱橫世界飲食舞台，贏得饕餮一族的掌聲，現先簡略介紹中式烹調的流程：

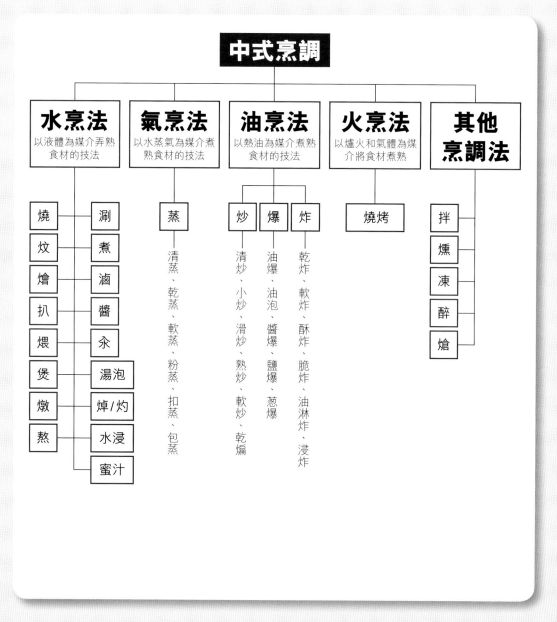

炸

清炸：

清炸是炸法之中最古老的一種，這種炸法是不掛糊的，特點是外焦內嫩有嚼口，方法是先將材料加工，醃漬，然後倒進滾油之中，用大火快速炸熟即可。

醃漬過程中有幾方面要留意，首先時間不宜過長，一般不超過一小時，可在炸前才醃，味道不妨淡些，因為經油炸後，物料會去失去部份水份，味料的味道更明顯。

醃料中的醬油亦不宜太多，甚至可以不用，否則炸時色澤會過深，不好看。食物炸好後，如果味道比較單一，可另跟調味料上桌，這是豐富味道層次。

乾炸：

乾炸是從清炸中分出來的一種技法，物料在醃漬後，撲上一層乾澱粉或水粉糊，再投入油鍋內炸，炸好的成品色澤較深，呈焦黃色，乾爽乾香，鬆脆味濃。

酥炸：

酥炸的原料必須是預先製好的熟料，由於預製的原料表面和內部都會含有較多的水份，因此在上糊和油炸前，都必須抹乾水份，否則不易掛糊，也很難掛勻，在油炸時亦容易脫糊，水和油會劇烈地爆起，發生危險。

如酥炸排骨，先以蛋液、生粉等調料醃漬，炸前再撲上乾粉，投入旺火熱油中，採用一次炸或兩次複炸成菜的技法。

軟炸：

是把醃過的材料，再沾上麵糊，放下油鍋炸成，炸油應是原料的5-6倍，要蓋過用料。油在鍋內燒至八成熱時改為小火，然後迅速把用料投入，全部下鍋後，再開大火，炸至表面呈金黃色，成品外脆內軟，而用料必須無骨。

淋炸：

把材料吊起，利用高溫油反覆地從上向下澆潑，還要從不同的角度澆遍全身，目的是使整料均勻受熱，熟度一致，這樣做法，行內習慣稱之為"淋油"。

潑炸：

將細嫩的鮮料，如稚嫩體小的子雞、乳鴿、鵪鶉或魚、蝦、蟹等物料，經醃漬後放在漏勺中以七成熱以上的油邊澆邊轉動漏勺，使厚料在轉動移位中能均勻受熱，熟度一致，這種做法，行內習慣稱為"潑油"。

而在使用潑油炸食物時，所要注意一些問題：

* 如果炸料是生的話，必須要有充足油量，另外火要大，油溫一定要保持七成以上，但不能超過八成，既要炸脆、上色，又要內外成熟均勻，澆潑次數較多。

* 如果炸料是熟的話，炸的時間便較短，而成品也以外皮略脆、顏色夠深為佳。

捲炸、包炸：

將鮮嫩易熟而無骨的原料，加工切成細小的如塊、片、條或絲的形狀，或是將原料剁爛成茸泥狀的餡料，經過醃漬後，以外皮包捲，放入適當的油溫中炸鬆脆，便是捲炸或包炸。

捲或包炸的外皮大致相同，如：腐皮、班戟皮、蛋皮、網油、威化紙(即米紙)，也可選用蔬菜葉或薄肉片、魚片等。而無論甚麼皮都好，都要包緊，餡料要均勻，如要掛糊或撒麵包糠，也要確保牢密平均，這樣才能保持形狀，油亦不會瀉入餡內。

但捲炸和包炸還是有分別的，就是形狀不同。包成捲筒狀的稱之為"卷"，包成長方形、方形、三角形的就要稱做包，如用玉扣紙、玻璃紙包裹餡料來炸的，便稱作"紙包炸"了。

前兩者可連皮進食，後者吃時要去掉紙張，故現在比較少人用，是較古老的包炸法。

至於火力控制，便要視乎所用外皮的耐熱程度而定，油溫方面，大致都在5-6成或7-8成之間。

吉列炸：

將物料蘸粉，拖蛋漿，再滾上麵包糠，然後炸，亦可把乾粉(生粉、粟粉、麵粉其中一款均可)加入蛋液中調勻，再上麵包糠，然後炸之。前者工序較多，後者簡略了些。

油浸

將原料放入高溫熱油中，使表面受到油的高溫，迅速形成了薄膜，隨即端鍋離火，或將火減慢，使油溫降低至七成熱以下，保持定低溫，使熱能慢慢滲入原料內部，產生變性、脫水、由生變熟時，即可撈出。

這種技法又稱浸炸，多用於魚類，由於質感外表略脆，而內鮮嫩，故烹製較細小的家禽也很適合，如仔雞、乳鴿及鵪鶉等。

烹

將切配好的原料，經醃漬後，撲上乾粉，放入大量熱油中，炸至金黃酥脆，盛起瀝油，隨即爆香料頭及配料，將預備好的調味汁料注入鍋中，迅速把原料回鍋，加入汁料，成品便外脆肉嫩，而又能被調味汁料全面包裹。

熘

熘的技法有三種，分為焦熘、滑熘和軟熘。其中以焦熘和烹法最為相似，兩者都是由炸法演變而來。

"熘"是將加工好的原料，經過不同方法，如蒸、煮、氽、焯、燙、炸等初步處理，然後再回鍋中，以短時間加熱，使芡汁能快速裹勻原料成菜。

由於熘的預製方法較多樣化，便形成了菜餚的多種質感、熘汁的多種調味出現，其味型除鮮鹹及酸甜外，更增加了麻辣、鹹甜、微酸(醋熘)和魚香等等。

煎

煎是一次性的加熱成菜，通常只用一種主料，以適量的油，中小的火，將原料在受熱的過程中，形成了香脆金黃的表層，鎖住了內部的肉汁和鮮味。為防止黏，放物料煎製前，紅鑊熱油是必要的條件，為使原料能易成熟，一般都切割成扁平或塊狀，厚度適中，或將原料剁成泥茸，做成扁平狀，再進行煎製。

貼

一面為貼，兩面為煎。以紅鑊慢火，放下適量的油，將食物放下，把一面煎成金黃色，如煎太陽蛋方式便是"貼"。貼法調味因菜而異，可先醃漬原料，或加入調味料再煎，或待食用時方加味汁，其特色是一面金黃酥脆，一面軟嫩清鮮。

塌

塌將原料兩面粘上糊漿，平舖入鍋，以適量油中慢火煎，呈金黃色，添湯加輔料、調味，使原料吸收入味而成菜，在餐飲業中，"煎、貼、塌"三法視同一類，都是以易熟細嫩的原料為主。

塌菜不宜多添湯，當湯快乾時，即加少許芡料出勺，如塌菜不勾芡就叫"貼"。

炒

"炒"是將主料泡油，泡至六七成熟盛起，再以適量油，以猛火將料頭和配料爆香，加入主料，即潷酒、加味打芡，炒勻便成，手法要快。

芡要掛而不瀉，即炒菜吃完碟上沒有芡汁為佳，芡汁多便變成扒菜。

炒菜要講火侯與鑊氣，而鑊氣的關鍵在於爆香料頭和潷酒之間所發出的香氣。

炒菜的特點是時間短、火候急、汁水少，大體可分為生炒、滑炒、軟炒。

生炒：

生炒又叫煸炒和小炒。

生炒的做法是，原料不用醃漬，不上漿，以猛火熱油，在短時間內加熱，充分體現原料的質地特點和鮮味，具有清新脆嫩的效果。

滑炒：

以肉、魚、蝦作主料，將主料以蛋清和淀粉拌勻，泡嫩油，然後再以少量油起鑊，爆香料頭，再將泡過油之主料與輔料放入，加進調味炒勻，以少許淀粉勾芡便可出勺。

軟炒：

液體原料加調味、輔料等拌勻，以油（或水）炒製凝固成菜。這種技法所用的主料都是液體，如牛奶、蛋液等，其輔料都是切碎或剁蓉泥狀的。

爆

"爆"利用旺火沸油或沸水將切成小塊形的原料進行瞬間加熱,再放入有少許熱油的鍋內,加調味汁成菜的技法。

油爆:

用多量沸油將小塊形原料進行瞬間加熱(加熱前大多先用沸水焯燙片刻)後,再加入芡汁攪拌成菜的技法。

醬爆:

將材料切割成大小、厚薄、粗細一致,且有的需剞花刀,先上漿或掛糊後泡油處理。調味料中一定要有甜麵醬,再加材料及其他綜合調味料爆炒,特色:鮮、爽、嫩、脆、芡汁掛而不瀉。

拔絲

"拔絲"是中國甜品製作中最具特色的技法,將僅熟的主料處理好,經掛糊以油炸脆後,放入熬好糖漿的鍋內拌勻,使主料能迅速包裹着一層有黏性而沒有完全凝固的膠狀糖漿,這種糖漿既能黏住主料,又能拔出細長的糖絲,故名"拔絲"。

成菜後以筷子夾出主料,拔出糖絲後,立即放在冰水裏一浸,黏在主料表層的糖漿便會立即凝固成一層晶瑩剔透,明亮鬆脆的薄殼,香甜可口,拔絲甜品適宜熱食。

掛霜

"掛霜"又稱"反沙"。

霜是由糖溶液經受熱重新結晶而成,糖溶液因受熱不斷出現結晶,晶粒聚集變大,當停止受熱,溫度下降,晶粒便會由大變小而分散開,最後成為相等於砂糖的小晶粒,形如白霜。將加工好的熟料放入熬好的糖漿熱鍋內,拌勻糖漿,立即熄火降溫,使成菜表面泛起白霜。

高麗

"高麗"即是把蛋白打成厚忌廉狀,拌入粟粉和麵粉成蛋白糊,以慢火浸炸成微淺金黃色澤。成菜後灑上糖霜,入口鬆軟可口。

甘脆濃腴的油烹法

炸
清炸
乾炸
酥炸
軟炸
脆炸
淋炸
捲、包炸
吉列炸
油浸
烹
熘
焦熘
煎
貼
塌
炒
生炒
滑炒
軟炒
爆
拔絲
掛霜
高麗

印尼香炸雞翼

成菜特色

焦黃甘香，肉嫩可口。

⏱ 20分鐘　👤 4~5人

烹調要點

1　另一種的炸雞方法，先將雞件放在炸籬上，不斷淋上滾油至呈金黃即可。這樣做可避免雞件黏鑊，不過比較費時。

2　如果不喜歡用雞翼，可用½隻雞來取代。

3　印尼人用芫荽粉和香茅粉的份量較多，約1~2茶匙，我認為這會影響雞的鮮味，加上很多香港人不習慣太濃烈的香料味，所以在此已稍加調校。

4　芫荽粉可改用芫荽籽，份量便要1茶匙，而香茅粉可用½支香茅代替，舂爛便可。

每菜一食材

黃薑

黃薑屬薑類的一種，是咖喱食物的主要香料，具調色調味的作用，含天然獨特的味道，顏色金黃。市面有鮮貨和研磨後的粉末產品，前者的味道不及乾貨濃烈集中，但卻有清新香味，粉末狀的黃薑就味道集中，鮮薑香味卻不明顯。

單一料：雞翼

材料：	醃料：	調味料：
雞全翼5隻	黃薑2片	胡椒粉¼茶匙
	石栗2粒	鹽1½茶匙
	乾葱2粒	芫荽粉½茶匙
	蒜頭2粒	香茅粉½茶匙
	椰汁¼杯	黃砂糖1湯匙
		雞粉1茶匙
		李派林喼汁½茶匙

做法：

1 雞翼洗淨，抹乾。

2 把醃料用攪拌機攪爛成漿。

3 將雞翼置於盆中，加入醃料和調味料拌勻，醃約1小時後備用。

4 油燒熱，將雞翼放下炸成金黃色，瀝去油分至雞件略為乾身即成。

炸魚餅

成菜特色

色澤金黃悦目、鮮嫩、爽滑富彈性。

⏱ 15分鐘　👤 4~6人

烹調要點

1. 蘸汁料適合蘸炸魚、蝦及其他海鮮食用。

2. 手剁或機攪碎的魚肉，各有優劣，前者的魚肉茸會比較多細骨和粒子較大，不及機攪幼細，但勝在魚肉的粗纖維仍在，肉丸或肉餅會富彈性和嚼口；後者肉質細緻，不易有碎骨，但肉纖維容易被破壞，彈力不及手剁的魚肉高。另一種方法，便是用匙或小刀刮出魚肉，兼取手剁和機攪的優點，但魚肉不能完全被刮去，浪費了魚肉。

3. 魚肉要順方向攪，邊攪邊加入適量水分，才可攪至起膠而有彈性。如有時間，最好放入雪櫃冰凍片刻才用。倘若想節省時間，可在魚檔購買已攪好的魚膠使用。

每菜一食材

青豆角

豆角又叫豇豆，盛產於夏天，豆莢如管狀，質脆而身軟，微腥，起鑊時加入葱粒，以猛火爆炒，可辟除腥味。

主料：魚肉；輔料：青豆角

材料：	調味料：	蘸汁料：
魚肉240克	辣椒膏1湯匙	魚露、椰糖各2湯匙
青豆角20克	醬油、鹽、糖、雞粉各¼茶匙	酸子汁3湯匙
雞蛋½隻	生粉½湯匙	蒜茸、紅辣椒碎、乾葱茸各1茶匙
檸檬葉1茶匙	油½茶匙	花生碎適量
	胡椒粉、麻油少許	

做法：

1 魚肉攪爛；青豆角切細粒；檸檬葉切碎；再將它們與雞蛋、魚肉同置於盆中，加入調味料，攪至起膠。

2 手沾點油，捏成10個小魚餅。

3 將魚露、椰糖及酸子汁煮沸片刻，待凍後加入其他蘸汁料拌勻。

4 燒熱油，以中火將魚餅炸成金黃色，食時伴以蘸汁料。

紫竹仙蹤

味鮮適口、紫菜清香獨特、油而不膩。

清炸

⏱ 40分鐘　👤 4~6人

烹調要點

1 包好的腐皮要上籠蒸30分鐘，目的是蒸發多餘水份，使其不會太濕身，以避免油炸困難，腐皮蒸過後待凍，再放入雪櫃，待翌日才拿出來炸，效果會更理想。

2 腐皮又稱為鮮竹皮，具柔軟和韌度，不易碎裂，以前在豆腐店或粉麵店有售，現在則有冰鮮或急凍貨，在一般南貨店或凍肉店也有售賣。未使用前，必須放在膠袋中封密保存，確保其濕度和柔軟，失掉了水份的腐皮會變乾脆，容易折斷，做不到預期效果。

每菜一食材

紫菜

以表面光滑滋潤、紫褐或紫紅色，有光澤、不黯淡為佳。

坊間的紫菜主要來自日本和韓國，日本的紫菜片會比較油潤和柔軟，味道濃郁，但沒有太重鹽份；相對地，韓國紫菜就略嫌粗糙，鹹味很重，不過亦有一些不帶鹽份。最新一種紫菜為岩燒紫菜，比較鬆脆乾爽。

主料：腐皮；輔料：紫菜

材料：
腐皮1斤（約10件）
紫菜10張
薑4兩（160克）
清水6量杯
牛油紙一張

調味料：
鹽2湯匙
味之素2茶匙

做法：

1 把薑洗淨，磨成薑汁。

2 將6量杯清水置煲中，加入調味料和薑汁，調和。

3 將腐皮摺成三角形。

4 把滾湯淋下，將多餘的湯水倒出。

5 腐皮平放，在中心加入1張紫菜，再摺成長方塊。

6 以牛油紙包好，放入蒸籠裏，用大火蒸30分鐘取出，待涼，放進冰箱。

7 食用前，取出炸成金黃即可。

香蒜椒鹽明蝦

光亮艷紅、蒜香飄逸。

✓ 10分鐘　🍴 4~6人

烹調要點

1 炸蝦時，不宜撲太多乾粉，因為油炸時容易變焦，而且會失去油香，炸得也不透。而炸好後，最好濾一下油，因為會有很多炸油沉澱物積在油裏。

2 蝦含水份比較重，特別是蝦腦含水份特別多，亦是腥味的來源，必須把部份蝦頭剪去，避免油炸時高溫令蝦中水份膨脹而爆裂，熱油四濺，弄傷皮膚。

每菜一食材

蒜頭

蒜頭屬葱科植物，又稱大蒜，它與洋葱、青葱、韭菜同一家族。大蒜含有豐富的蛋質、脂肪、維生素A、B_1、C、醣類及少量的鐵、磷、大蒜素、大蒜油、增精素等，對人體健康有助益。蒜頭經油炸後會散發一種香醇獨特香味，用作提鮮調味，十分普遍，廣東人做菜就缺不了它們。有些藥膳食譜如肉骨茶、珧柱甫、啫啫煲就加入大蒜，添加香味，以及用養生方法進補。

單一料：蝦

材料：

大隻中蝦1斤（600克）
紅辣椒仔1隻（切碎粒）
蒜茸2茶匙
花椒1茶匙（白鑊炒香）
鹽2茶匙（白鑊炒香）
生粉適量

醃料：

味之素½茶匙
胡椒粉少許
麻油少許

做法：

1 將蝦剪去鬚腳及剔除蝦腸，洗淨吸乾水分。

2 白鑊炒香花椒，加入鹽，以慢火炒至變色盛起。

3 將炒香之花椒鹽篩過，加入醃料中，與蝦拌勻，醃約15分鐘。

4 泡油前撲上適量乾生粉。

5 燒滾油，將蝦放下，以大火炸2分鐘盛起。

6 燒油少許，爆香蒜茸、紅椒碎，將蝦回鑊快速兜勻上碟。

果香陳皮骨

成菜特色

外酥裏嫩、陳香溢齒，佐酒佳餚。

☑ 20分鐘　👤 4~6人

烹調要點

1　如想炸出來的排骨更軟嫩可口，則需在醃骨前，先以少許蘇打食粉拌勻，醃約一小時左右，然後用清水沖乾淨，才以調味料醃骨。

2　陳皮是新會柑皮經剝柑肉後置北風風乾和太陽下晾曬的乾燥製品，年份越久，味道越濃，色澤也越深如濃茶。選用老年陳皮，份量宜少不宜多，因為其味道濃郁，很少份量已很濃烈，但陳皮的年份越淺，就要多用一點了。

3　話梅是現時潮流美食的配料，它的味道鹹中帶甜，甜中卻有獨特醃漬鹹味，用作排骨提味，十分匹配，但使用前必須確認其鹹甜度而決定用量多少，因為某些品種會帶甜口，而又些卻偏鹹，所以必須試味，避免過鹹或過甜。

每菜一食材

排骨

由五花肉上方取下來的小排骨，肥瘦均勻質地很嫩，適合製作乾炸排骨，京都排骨，糖醋排骨等用途。

單一料：排骨

材料：

一字排骨½斤（300克）
新會陳皮（浸透）½個
話梅（起肉）2粒
蛋黃1個
生粉適量

醃料：

蠔油1湯匙
醬油1湯匙
雞粉½茶匙
糖1茶匙
玫瑰露酒1茶匙

做 法 :

1 排骨斬件,陳皮及話梅肉浸透,剁碎。

2 將以上材料置盤中,加入醃料及蛋黃拌勻。

3 炸前撲上適量乾生粉。

4 放八成滾油中,炸至金黃熟透,即成。

香酥鴨

色澤金黃、外脆內酥、鴨肉香郁、肥而不膩。

☑ 4小時　　👤 6~8人

烹調要點

1 鴨的醃漬時間不宜過長，否則會太鹹。上籠必須將鴨蒸至酥爛，才能保持酥香之特色。

2 鴨肉蒸好後，必須放涼，方可撲生粉油炸，才容易炸脆。

3 油炸鴨肉前，放入乾葱片浸炸，讓其香味經高溫而釋放出來，鴨肉在油炸過程中會吸入葱油的香味，可減輕鴨的羶味，並提升鴨香味道。

4 剛炸好的鴨肉不宜立刻斬切，容易弄碎而不完整，最好攤涼片刻，切口才完美。

每菜一食材

鴨

鴨是飲食業中常用的原料。鴨肉味甘性平，主要成分為蛋白質、碳火化合物、脂肪、硫氨素、核黃素、磷、鐵、鈣等，有滋陰補血、利水養胃之功效。鴨肉雖然有益，但為多脂肪食物，不宜多吃，尤其是肥胖、動脈硬化者應少食為佳。

單一料：鴨

材料：	醃料（炒香）：	調味：
光鴨1隻	花椒2湯匙	紹酒¼杯
乾葱片40克	幼鹽60克	葱2條
生粉適量		薑片40克
		茴香1湯匙
		桂皮2片
		陳皮1角，浸透

做 法：

1 鴨去肺，洗淨，瀝乾。

2 用炒香之花椒和鹽擦勻全身內外，醃半天。

3 取出已醃好之整鴨，以清水沖淨，置碟上，加入調味，放入蒸籠，以大火蒸至熟爛。

4 取出，瀝去汁水，棄去葱、薑、桂皮等香料，使其自然冷卻，撲上適量生粉。

5 燒熱多量油，放下乾葱片，以慢火炸至金黃盛起，使油中留有香味。

6 放下已凍之整鴨，以八成滾油炸至呈金黃色，盛起瀝油，稍涼斬件，整齊地排回原狀，上碟。

河仙豆腐

成菜特色

色澤金黃、外脆肉軟、甘香可口。

⏱ 20~25分鐘 👤 4~6人

烹調要點

1. 河仙豆腐製法與太史豆腐相似。豆腐蒸熟後，必須待至全凍，方可切件炸之，不然會鬆散。

2. 選用臘腸，按肥瘦的比例3:7最理想，因為太瘦的臘腸會很韌，相反地肥肉比例太多，就吃得滿口肥油，肥膩又不健康。至於該臘腸採用真正豬腸衣做臘腸，就要用沸水微燙片刻，它才不會韌。

3. 冬菇浸泡後，壓乾水份，用點生粉撈洗，再用清沖洗，可把冬菇的泥霉味去掉。

4. 用刀剁馬蹄茸，不及用刀背拍碎而脆口，還會因鐵器或金屬令馬蹄容易氧化，所以可把馬蹄放進膠袋內，用木棍椿碎，原汁原味。

5. 木板豆腐味道濃郁，質感不夠細緻，容易變壞和變酸臭。盒裝豆腐，味道清淡，勝在質感細緻，容易貯藏，不易變質。按個人喜好而選用豆腐。

每菜一食材

馬蹄

原名荸薺，種於水中，可生吃，也可加工製成粉供烹調使用，以廣西桂林或廣州泮塘出產為上品。

有些馬蹄的水份不足，澱粉很重，味道濃郁，比較乾硬，但嚼口鬆脆，卻帶有渣滓；有些馬蹄則含豐富水份，沒有渣滓，味道比較清淡。爛了的馬蹄會出現黃色，輕按軟腍腍，宜取硬實，扁身和沒有異味的馬蹄，品質較優。

主料：豆腐、馬蹄；**輔料**：冬菇、臘腸、蛋白

材料：

豆腐300克	蛋白2隻
去皮馬蹄200克	葱粒1條
冬菇2隻	生粉½杯
臘腸20克	味椒鹽適量

調味料：

鹽、雞粉各1茶匙
糖½茶匙
麻油、胡椒粉各少許

做法：

1 將馬蹄洗淨，磨成茸。

2 把冬菇和臘腸切成幼粒。

3 加入蛋白、豆腐、葱、生粉及調味料等，攪勻成醬。

4 將攪好之醬料倒進一個已墊上牛油紙之圓盤內，用手壓平。

5 用大火蒸15分鐘，取出，待涼後切成角形，並滾上一層薄薄的生粉。

6 燒滾油，把豆腐炸成金黃，撈起瀝油，灑上味椒鹽即可。

百花釀蟹箝

色澤金光明亮、彈牙爽口、筵席佳餚。

⏱ 15分鐘　👤 4~6人

烹調要點

1 如要蝦膠熟後夠爽口而又不易鬆散，關鍵在於攪剁前，要將蝦肉徹底抹乾，而刀與砧板也要保持乾淨。

2 攪拌時要注意力度及保持一致方向，攪拌好的蝦膠要放雪櫃內冷凍，最少2小時才可取出烹製。

3 蝦膠用的蝦以硬殼海中蝦最好，因為其肉質結實，膠質豐富，打出來的蝦膠會結實有鮮甜味道。值得一提，雪蝦比較適合做蝦膠，過於新鮮的蝦不易脫殼，要是鮮蝦太鮮，就要放冰箱雪1~2小時，讓其易於離殼。

每菜一食材

蝦

質好的蝦隻，頭部要完整，身體要保持原有的彎曲度，色澤青綠或青白，蝦殼發亮。

不是任何蝦皆可以攪打蝦膠，只有含膠量豐富的海蝦，如海中蝦、花竹蝦、黃蝦等最宜做蝦膠。

主料：蝦肉；輔料：蟹箝

材料：	調味料：
蟹箝8隻	生粉2茶匙
蝦肉½斤（300克）	鹽½茶匙
生粉適量	蛋白1湯匙
	麻油¼茶匙
	胡椒粉少許

做法：

1 蝦仁洗淨，去腸，用布吸乾，放攪拌機攪爛。

2 蝦肉加入調味打成蝦膠放雪櫃。

3 用時取出釀在蟹箝上。

4 黏上生粉，放油中炸成金黃色即可盛起。

酥炸桂林牛肉丸

外酥內嫩、鮮香可口。

⏱ 10分鐘　👤 4~6人

烹調要點

1 將攪好之肉丸，放1粒在冰水中來測試，若肉丸浮起則攪撻成功，可以烹製。

2 在牛肉下鍋前，與少許蘇打食粉拌和，能使其口感鬆嫩，而醃肉時加些熟油拌勻，除可防止泡油時結塊外，還可去腥增香。

3 牛肉丸需要混合肥肉融和，方可令肉質肥美帶肉汁，不會太乾硬，味如嚼蠟，不好吃。

4 肉丸沾上蛋白糊，再滾上生粉才油炸，可保肉汁不流失。

5 做肉丸的牛肉不要有太多肉筋膜，否則其會很粗糙而不幼滑。

每菜一食材

牛肉

牛肉屬紅肉，味道濃，不同部位的牛肉要懂得配合烹調方法，方能充份發揮它的美味效果。帶有脂肪的肉質適合煎、炒、焗、炸、焯、燙，如牛肩肉、肥牛肉、牛柳等。少點脂肪但肉纖維較多的牛腱(牛脹)、牛腩等就適合炆、燉、煮、煲、熬等。

炆、燉牛肉時，可將約4-5條稻草洗淨，打成結投入鍋內，或加紅棗幾粒，均能使肉起酥香作用，不妨一試。

主料：牛肉；**輔料**：馬蹄、肥肉

材料：	蛋白糊料：	調味：
牛肉6兩(240克)	蛋白1隻	蠔油2茶匙
肥肉2兩(80克)	生粉2湯匙	醬油1茶匙
馬蹄肉2個，剁碎		糖1茶匙
辣椒仔1隻(去籽，切碎粒)		生粉1湯匙
生粉適量(後下)		蛋白1湯匙
		麻油少量
		胡椒粉少量

做 法 :

1 將牛肉和肥肉剁碎,放進碗裏。

2 除生粉外,加入材料及調味,攪至起膠。

3 做成肉丸,沾上蛋白糊,再上乾生粉。

4 放入中火熱油中,炸成金黃色,即成。可以蘸汁同食。

香芒威化卷

色澤金黃、香甜鬆脆。

酥炸

☑ 25分鐘　👤 4~6人

烹調要點

1 要使麵衣炸後入口酥脆爽口，務必在油炸前才做脆漿，這樣才不會產生黏性，脆漿做好後，靜置最好不超過15分鐘。

2 以生果製作油炸食品，最適合是即包即炸和即食，否則容易出水而不香脆。

3 生果本身含有天然果酸和豐富果糖，這些成份會受高溫而產生化學反應變酸，建議用快速加熱的方法處理，避免過熱變酸，還要選用比較熟的芒果入饌，製品的味道才會集中而仍保持蜜味。

4 威化紙遇水會容易融化，所以不要讓其沾水。

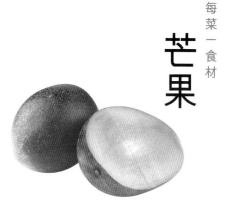

每菜一食材

芒果

香甜的芒果是很受人們歡迎的水果，品種繁多。常見的有來自菲律賓的呂宋芒，而象牙芒更有"芒果之王"的稱號。至於橢圓形的"白花芒"，皮色黃澄，則有開胃作用。而大個的"蘋果芒"，皮色綠中帶紅，質感較粗糙。最小的"豬腰芒"形似豬腰，核薄如紙，甜美可口。其他品種還有甜中帶酸的"枇杷芒"，和糖漬後才好吃的青芒等。

主料：芒果

材料：	脆漿料：	蛋白糊：
芒果1個	炸粉½量杯	蛋白1隻
香蕉1隻	吉士粉1湯匙	生粉2湯匙
沙律醬2湯匙	凍水½量杯	
威化紙適量	油3湯匙	

做法：

1 芒果去皮，去核，切粗條，約3吋 x ½吋 x ½吋。

2 香蕉去皮，切粗條，約3吋 x ½吋 x ½吋。

3 脆漿料預先開勻，15分鐘後備用。

4 把芒果、香蕉加入沙律醬拌勻，以威化紙包好，用少許蛋白糊，加上脆漿，放入沸油
中炸脆、顏色成金黃。

燒雁鵝

成菜特色

甘香酥脆、鮮美可口。

⏱ 20分鐘　👤 4~6人

烹調要點

1 即食滷水鵝方便即時做這菜，沒有家庭味道，商業味道比較濃，如果有時間可以自己滷製，味道會較適合自己需要。

2 自製滷水，變化很多，台式風味可到台灣食品店購買台式滷包，會含有紅麴米成份，所以有時會偏嫣紅色；港式滷水料可到藥材舖購買滷水盆，桂皮、草果、甘草、八角、花椒、香葉、沙薑、陳皮等再配以冰糖和老抽，味道鮮淡，別有一番滋味。要時又想自做滷汁，卻花不起時間熬煮，可買現成方便滷汁，再加入部份香料，增強其味道如八角、花椒、辣椒等，必需在滷汁完成後，灑點玫瑰露酒添香。

3 自己動手滷鵝，要揀肥瘦適中的鵝，太肥的鵝會在烹煮過程，不斷浮出肥油，肉少油多，滿口肥膩，太瘦的鵝又吃不到預期風味，宜採用3~4斤重的鵝，無論味道和肉質符合這菜的需要。

4 可用生菜墊底，或將黏着鵝肉的骨斬件，加入餘下之蛋糊拌勻，再上生粉炸香放碟上，把炸好之雁鵝排在鵝骨上。

每菜一食材

鵝

鵝是家禽中體型較大的種類，與雞鴨相比，肉質較粗，且有腥味，作為烹調原料，應用也不如雞鴨廣泛，主要用於烤、醬、滷、炖、炆等技法處理，以小火長時間加熱烹之。此菜餚亦可以鴨代之。

主料：鵝

材料：
滷水鵝1隻
酸甜辣醬或桔油適量
生粉適量

蛋糊：
雞蛋1隻
生粉3湯匙

做 法：

1 將滷水鵝起肉。

2 將蛋糊材料調勻，然後均勻地塗在鵝皮和肉上。

3 撲上適量乾生粉。

4 燒油至八成熱，小心放下整塊鵝肉，炸成金黃色，盛起瀝油。

5 將鵝肉斬件，排在碟上即成。

麵拖魚條

成菜特色

外脆內軟，味甘香。

⏲ 10分鐘 　👤 4~6人

烹調要點

1 依據縱直紋理切割魚肉，才能保持完整而不碎裂，相反地以橫紋切割，魚肉易斷不完整，油炸後變碎。

2 脆漿料除了用炸粉外，還添加了吉士粉和糯米粉，前者有添香提味的功效，還可令脆皮增加金黃色澤，後者則保持脆漿的脆中帶糯軟的效果，不會太脆硬而易碎。

3 用冰水調漿，會令脆漿的粉料筋性容易結網，弄出來的脆皮更甘香鬆脆。

4 凡是用作炸食之魚肉，醃製時不用放薑汁，因檸汁除可去腥外，更有增香之作用。

每菜一食材

苔條

又稱苔菜，真正的名稱是滸苔，其學名 Enteromorpha prolifera，屬綠藻科，它的主枝明顯，分枝細長，長可達1米，常生長在潮間帶岩石上或石沼中。它的分佈範圍甚廣，全世界各地海洋、半鹹水域或江河，或泥沙灘的石礫上，有時也可附生在大型海藻的藻體上。其含碳水化合物、蛋白質、粗纖維及礦物質、脂肪和維他命，屬高蛋白、高膳食纖維、低脂肪、低能量，且富含礦物質和維他命的天然理想營養食品的原料。

主料：魚肉；輔料：苔條

材料：	脆漿料：	醃料：
白肉魚（任何魚類均可）300克	炸粉½杯	雞粉1茶匙
苔條20克	吉士粉1湯匙	檸汁2茶匙
	糯米粉2茶匙	蛋白1湯匙
	凍水½杯	生粉1茶匙
	油3湯匙	熟油1湯匙

做法：

1 用清水洗淨苔條，然後用布吸乾，放入高溫油中，炸20秒，盛起壓碎備用。

2 把魚肉切成粗條狀，以醃料拌勻，醃約15分鐘備用。

3 脆漿料開勻，加入壓碎之苔菜拌透，20分鐘後備用。

4 注油入鍋中，燒至八成熱。

5 把魚條上脆漿，投入油中，炸硬，至色澤金黃即可。

脆炸子雞

色澤金紅、皮脆肉嫩。

☑ 40~45分鐘 👤 2~4人

烹調要點

1 白滷水料與清水熬煮15分鐘便足夠，因為子雞的肉味比較清淡，不能以太強烈的滷汁處理，所以滷汁有點香料味道便可。

2 把子雞放滷汁中煮滾便熄火，滷汁的熱力足夠令雞肉變緊熟而不變粗，方便作後期油炸的加工過程。

3 上皮料的麥芽糖能令雞皮風乾後緊縮，容易保持雞皮的收緊，當油炸時不會因冷縮熱脹而令雞皮容易爆裂。至於浙醋則是令雞皮油炸後會變赤紅，色澤鮮艷吸引。

4 上皮料淋在雞身時要均勻，否則油炸時雞皮的色澤會不均勻，影響到賣相。

5 淋油時盡量四週均勻地淋澆，才能保持雞身色澤鮮明而均，不會偏色。

每菜一食材

子雞

又稱"春雞"，是尚未成熟的小雞。肉質鮮嫩，味道清淡，一般是急凍貨，但高級食材店會有冰鮮的空運貨售賣。這些著名貨源有來自丹麥、法國、美國等，外國人不喜看到雞頭，所以他們會把雞頭去掉。子雞除了油炸、釀焗都是外國喜愛的菜式，不過韓國人也愛把鮮人參、薏米、紅棗和蓮子等釀入肚內，以中火燉煮至爛熟，味道鮮美，屬於韓國的著名菜式。

單一料：雞；配料：蝦片

材料：	上皮料：	白滷水料：	
光雞1隻(1斤12兩/1050克)	玫瑰露酒1湯匙	八角4粒	白胡椒1茶匙
蝦片適量，炸脆	浙醋2湯匙	丁香1茶匙	薑2片
	麥芽糖1湯匙	桂皮10克	乾葱頭3粒
	生粉1湯匙	甘草2片	鹽80克
		陳皮1角	糖60克
		沙薑20克	味粉1茶匙
		香葉2片	清水2-3公升

做法：

1 把白滷水料置煲中，注入清水煮沸，以中火將滷水煮約15分鐘。

2 將雞放入煮滾之滷水中，熄火浸約15分鐘，取出待凍備用。

3 將上皮料之玫瑰露酒、浙醋和麥芽糖煮熱，待凍，加入生粉調勻，然後在雞身淋約2-3
  次，使其均勻。

4 將雞吊在當風處約2-3小時，吹乾。

5 把油燒至七成熱，先將漏勺，放入油中燒熱片刻，然後將雞放在漏勺中，用潑油法，
  把子雞澆潑熟，成金紅色。

6 趁熱皮脆時，以利刀迅速將雞斬件上碟即成。

紅燒乳鴿

色澤金紅、皮脆骨香、肉嫩。

⏱ 30分鐘　👤 4~6人

烹調要點

1 乳鴿的眼睛含有水份，所以生炸乳鴿就要把其眼睛刺破，防止因受熱時眼球內的水份膨脹爆破，濺油弄傷自己。

2 當乳鴿浸泡時其體內的雜質或血水會因受熱變成污物黏附在鴿身，所以上皮前必須沖洗乾淨。

3 乳鴿的皮油炸後酥脆，必須把乳鴿上皮後徹底風乾，否則雞皮炸不透徹，容易變軟回潮。

4 上皮後不要用手摸，並要獨立吊起，否則油炸後鴿皮的顏色會變得不均勻。

5 炸時要不停轉動鴿身，使其受熱均勻，成熟一致。

每菜一食材

乳鴿

選購入饌用的乳鴿，體型嬌小為佳。乳鴿雖多由人工飼養，但依然帶有野禽的味道，如要煲湯燉製，最好剖開脊背或把整隻乳鴿切成兩半，這樣才能容易把乳鴿的精華溶入湯中。另外還要加入瘦肉、雲腿來去其異味。如要燒、焗、浸、滷，最好加入玫瑰露酒醃漬，使其香味更加特出。

單一料：乳鴿

材料：	上皮料：	白滷水料：	白胡椒1茶匙
乳鴿2隻	玫瑰露酒1湯匙	八角4粒	薑2片
	浙醋2湯匙	丁香1茶匙	乾葱頭3粒
	麥芽糖1湯匙	桂皮10克	鹽80克
	生粉1湯匙	甘草2片	糖60克
		陳皮1角	味粉1茶匙
		沙薑20克	清水2-3公升
		香葉2片	

做法：

1 將乳鴿洗淨，去肺，放入煮滾的白滷水中，熄火。

2 泡約15分鐘後，取出，用熱水沖淨鴿皮上的油污。

3 瀝乾，以上皮料塗勻，掛在當風處吹乾。

4 把大量油燒至七成熱，轉慢火放入乳鴿，炸至鴿皮轉色，改用中火，將乳鴿炸至色澤金紅便成。食時用准鹽、喼汁蘸食。

印尼雞卷

色澤金黃、鬆軟可口。

☑ 30分鐘　👤 6~8人

烹調要點

1 皮要煎得夠薄才好包,用易潔煎鍋以慢火煎成,切勿煎成焦黃,麵包糠要選擇幼粒的那種。

2 煎好的薄餅皮,不要重疊一起,容易弄破,建議可用保鮮紙隔離薄餅皮,方便預先處理,隨時應用。

3 吉列菜式屬西方菜餚,把物料包裹後沾上適量麵糊,再滾上麵包糠,然後油炸,外脆內軟就是其特色。但市面上的麵包糠款式很多,美國的麵包糠有用梳打餅碎,或是乾燥了的麵包碎屑,研磨極幼細。日本麵包糖用乾燥的麵包,有粗研磨碎或中度研碎,強調製品的形態和質感。自製麵包糠則可把隔了一夜的方包,用攪拌機弄碎,亦可把梳打餅弄碎,兩者混合使用或是獨立使用均可。

每菜一食材

煙肉

煙肉,台灣用音譯稱之為"培根"。此物多數出現在西式菜餚及早餐之中。這種煙燻過的鹹豬肉,大多是用豬腩肉或腩條肉作材料,普遍做法是炸、煎或烤食味甘香可口,做點心餡料,加入了煙肉可更冶味。

主料:混合料,即主輔料份量相約

皮料:	吉列皮料:	餡料:	白汁料:
雞蛋2隻	雞蛋2隻	雞肉粒、白菌粒、	牛油1安士(約30克)
糖2湯匙	麵粉2湯匙	洋蔥粒各½杯	麵粉¼杯
鮮奶1杯(約250克)		雜菜粒¼杯	水½杯
麵粉2兩(約80克)	調味料:	煙肉粒1片量	花奶1湯匙
生油¾兩(約30克)	雞粉、糖各¼茶匙	熟蛋粒1隻量	鹽¼茶匙
	胡椒粉少許		雞粉1茶匙
	麵包糠適量		

做法：

1 燒熱油，炒香洋葱及煙肉，加入雞肉、雜菜、白菌等炒熟，下調味料、鹽及雞粉，拌匀，最後加入蛋粒略炒成餡料，盛起備用。

2 用牛油炒麵粉，其間徐徐加入清水，慢火攪匀至水加完，即下花奶及餡料拌匀，盛起，待涼備用。

3 將蛋、糖打匀後，再加入鮮奶，分次加入麵粉，打至起筋後，加入生油拌匀，煎成薄圓形；包入拌好的餡料，即成雞卷，備用。

4 將包好之雞卷沾上已拌的雞蛋及麵粉，再粘上麵包糠。

5 將雞卷放入滾油中，炸成色澤金黃時即成。

麻辣羊肉春卷

成菜特色

色澤金黃光亮、卷皮酥脆、餡心鮮嫩。

捲包
炸炸

☑ 10分鐘 👤 4~6人

烹調要點

1. 炸春卷要有耐性，由於包入餡料時會捲上幾層，所以表面看似金黃，但內層可能有濕氣而未能炸透，凍後便會回軟，故應多炸一點時間才取出。

2. 八成油溫落鑊，以中慢火炸之，炸好後盛起食物才能熄火。

每菜一食材

羊肉

羊肉性溫，含有豐富的蛋白質和維他命B群，營養價值相當高，中國人喜在冬天進補，其實一年四季都可以成為桌上的佳餚。

羊肉不同部位，有不同的烹調法，選擇肉質細嫩的羊肉或羊排較適合作炸春卷用。

主料：羊肉

材料：	蛋糊（拌勻）：	調味：	
免治羊肉150克	蛋黃1隻	辣椒粉¼茶匙	雞粉1茶匙
洋葱碎¼杯	生粉1湯匙	孜然粉½茶匙	糖½茶匙
春卷皮10塊		花椒粉¼茶匙	生粉1湯匙
		醬油1茶匙	麻油1茶匙

做法：

1 將羊肉、洋葱碎置盤中，加入調味攪透。

2 春卷皮攤開，加入適量餡料，兩邊摺入，向外捲起成長卷形，邊上抹少許蛋糊收口。

3 放入八成滾油中，炸至金黃脆身，即可進食。

脆香鳳肝卷

甘香酥脆、色澤金黃、外皮鬆脆、餡料鮮香滑如

⌛ 10分鐘　👤 4~6人

烹調要點

1. 鳳肝俗稱雞膶，屬雞的肝臟，質感柔軟，用新鮮貨，要小心挑出血管或雜質，亦可用罐裝外國貨，味道豐富集中，因為其質感太軟滑，小心弄碎。

2. 包裹餡料時，必須用點力度，捲捏結實一點，不能太鬆弛，否則油炸時會不成形，不似卷狀。

3. 蘸沾脆漿，不能太稠密，薄薄一層便可，否則脆漿的形狀很難控制，不夠平滑完美。

每菜一食材

腐皮

腐皮具有黃豆的營養價值，能補充體力，易為人體吸收。腐皮含有47%的蛋白質、豐富的脂肪質、鈣和熱量，是病後和產後的滋養食品。

主料：混合料，即主輔料份量相約

材料：

雞肝或豬肝120克，
　用薑汁醃透
冬菇2隻，浸透切絲
榨菜20克，切絲浸淡
西芹120克，切絲
溫室菜苗40克，切絲
腐皮2張，開4份
蒜頭1粒，切茸
乾葱1粒，切茸

調味：

雞粉½茶匙
蠔油1湯匙
糖1茶匙
五香粉½茶匙
麻油、胡椒粉少許
生粉1茶匙
水2湯匙

脆漿料：

自發粉50克
吉士粉½茶匙
糯米粉2茶匙
梳打粉¼茶匙
水⅓杯
油1湯匙

做法：

1 把脆漿料拌勻，放置一旁，待15分鐘才可用。

2 先將雞肝或豬肝整塊用水煮至剛剛熟，即取出切條。

3 燒熱油爆香乾蔥、蒜茸後，將冬菇絲、榨菜絲、西芹絲和溫室菜苗炒透後，再加入調味料拌勻，取出放涼後，用腐皮包成春卷。

4 燒滾油，將春卷沾上脆醬料後，放入炸至金黃色，取出斬件便可。

吉列鮮魷環

金光油亮、香脆可口。

☑ 10分鐘　👤 4~6人

烹調要點

1 魷魚在烹調和應用上與烏賊（又稱墨魚或花枝）基本相同。

2 製作菜餚較為講究刀工，否則造型和口味均欠佳，如要爆炒則宜多油，動作要快，若炒過頭，肉質會變老。

3 油炸的菜式，其油溫控制是成菜的關鍵，所以用油的品質會直接影響到成品的效果。因為不同的油類，其冒煙點會有不同，油炸食物可用花生油，耐熱度高又有香味，葡萄酒的耐熱度也不錯，味道清淡，勝在夠健康，至於芥花籽油或葵花籽油就適合炒、焯用。

每菜一食材

魷魚

活魷魚色澤鮮艷，肉呈淺粉，表面有些白霜，背部呈半透明，肉質肥厚，頭腕堅韌。新鮮活魷，肉色潔白，質柔軟，味鮮。

單一料：鮮魷

材料：	醃料：	蛋麵糊（調勻）：	醮醬料：
鮮魷魚2隻（約½斤，300克）	淡奶3湯匙	蛋2隻	千島醬適量
原味粟米片適量	味椒鹽½茶匙	麵粉2湯匙	
檸檬，裝飾用	雞粉1茶匙		
	胡椒粉少許		

製法：

1 把鮮魷魚去衣洗淨，橫切成魷魚圈，與醃料拌勻，15分鐘後備用。

2 壓碎粟米片，將鮮魷圈投入蛋麵糊中拌勻，然後轉放在粟米片中。

3 將鮮魷圈放在約220℃滾油中，炸成金黃色即成，食時可用千島汁點食。

吉列靈芝菇鮮蝦沙律

色澤金黃、鮮甜冶味、爽口。

⏱ 10分鐘　👤 4~6人

烹調要點

1. 哈密瓜較蜜瓜味道清淡，但肉質爽脆，比較適合雕刻用。

2. 可選用蜜瓜或皺紋瓜，隨你喜好而定。

3. 鮮蝦在上碟前才炸較為好些。

4. 哈密瓜盅可預先雕好，然後用保鮮紙包裹，放冰箱中貯藏。

每菜一食材

哈密瓜

哈密瓜，是甜瓜的一個變種，屬葫蘆科植物。風味獨特，味甘如密，香氣襲人，有"瓜中之王"的美譽。哈密瓜有180多個品種及類型，不同品種的瓜，其形態、顏色、皮紋也不一樣。市場上全年都有供應。

哈密瓜富有營養。據分析，哈密瓜含有糖分4.6%-15.8%，纖維素2.6%-6.7%，還有果酸、果膠物質、維他命A、B、C，菸酸及鈣、磷、鐵等礦物質。其中鐵的含量比雞肉多兩三倍，比牛奶高17倍。

主料：松茸、鮮蝦；**輔料：**肝醬、洋葱

材料：

靈芝菇300克
中蝦12隻
洋葱½個，切粒
肝醬(任可肝醬均可)40克
哈密瓜½個
威化紙適量
麵包糠適量
沙律醬適量

醃蝦料：

蛋白1茶匙
生粉¼茶匙
鹽⅛茶匙

蛋糊：

雞蛋2隻
生粉4湯匙

調味：

鹽¼茶匙
雞粉½茶匙

做法：

1　洗淨靈芝菇，汆水；鮮蝦去殼及腸，留尾起雙飛，以醃料拌勻，炸熟。

2　哈密瓜連皮雕成帶花邊的盅，備用。

3　以少許油爆香洋葱及靈芝菇，下調味，拌勻，盛起，瀝乾。

4　將一半洋葱、靈芝菇拌入肝醬，以威化紙包成春卷形，用少許蛋糊塗在收口處。

5　把包好之松茸卷拖蛋糊，再滾上麵包糠，放八成熱油中，炸成金黃色盛起。

6　將哈密瓜肉切丁，加入餘下之靈芝菇，和以沙律醬，放入哈密瓜盅內，再把炸好之鮮
　　蝦加入，裝盤上桌。

油浸香茅仔雞

成菜特色

油澤金黃，外皮略脆，肉質十分鮮嫩。

☑ 15~18分鐘　👤 4~6人

烹調要點

1 這種浸炸技法，與用沸水浸燙的"白切雞"基本相同。但油浸的雞總比水浸的更能保持原汁原味，且原料色澤更具光亮鮮艷。

2 要注意的是，當物料未放入前，油溫必定要燒至八成。

3 物料放下後，油溫減至七成，在浸炸這15分鐘內，油溫保持定六成至七成，便會製作出理想的佳餚。

每菜一食材

仔雞（小雛雞）

仔雞即小雛雞，別名"小筍雞"，一般生長期為2個月左右，重量約250克，肉質最嫩，但出肉少，適宜連骨製作菜餚。

單一料：仔雞

材料：	醃料：	點雞汁料：
仔雞1隻(400克)	鹽焗雞粉2茶匙	青檸½個，切小塊
	醬油1茶匙	辣椒仔1隻，切粒
	玫瑰露酒1茶匙	蒜茸1茶匙
	熟油1湯匙	魚露1湯匙
	香茅1支(切段，略拍)	糖2湯匙
	八角1粒	

做法：

1 雞洗淨，去肺，以醃料塗勻雞肚內外，再把香茅和八角塞進肚內，醃1小時備用。

2 取出香茅、八角，用吸水紙吸乾水份。

3 將雞吊在擋風處，吹乾約1小時。

4 燒熱多量油至八成熱，放下仔雞約2分鐘，隨即熄火，待浸約15分鐘後，即可取出，
　待涼後斬件上桌。

油浸香醋鯇魚尾

清雅醒目，魚肉細嫩，醒胃適口。

成菜特色

⏱ 10分鐘　👤 4~6人

烹調要點

1 魚肉雖然鮮美，但帶有令人討厭的腥味，要清除並不困難，一般在烹調魚類的過程中，加入適量料酒和少許醋就行了，生品成熟後，再放上葱、薑，潷下滾油，就可以大大減輕或全部將腥味清除。

2 任何魚味清淡的魚類都適合烹飪此菜。

3 不同品牌的香醋都適合，如果用意大利黑醋或西班牙黑醋，味道更好，用量也不需多，因為它的味道濃郁又清純。

每菜一食材

鯇魚

草魚俗稱鯇魚，嘴小鱗大，魚腸肥美，每一部份都可烹製出不同的佳餚。愛用鯇魚尾雖然幼骨比較多，但因其整天不停擺動，運動量足，肉質結實幼滑，只要小心一點挑骨，頗適合浸、煎、炒、炸、焗、炆、啫啫煲和薑葱焗等烹調方法。

單一料：鯇魚

材料：

鯇魚尾1條（480克）
嫩薑絲20克
葱絲20克

調味：

鎮江香醋2湯匙
醬油2湯匙
黃砂糖1½湯匙
鹽⅛茶匙
味粉¼茶匙
生粉½茶匙

做法：

1 魚尾洗淨抹乾。

2 放八成熱油中浸2分鐘，熄火，浸8分鐘盛起，瀝油放碟上。

3 調味拌勻煮沸，淋在魚面。

4 放上薑、葱絲，再淋上滾油即成。

白汁海上鮮

成菜特色

顏色悅目，魚肉滑嫩，味鮮適口。

⏱ 10分鐘　👤 4~6人

烹調要點

1. 製作鮮奶時，油溫不宜高，鮮奶下油鍋時，應邊攪動邊注入。

2. 菜譜中的做法是中式簡易白汁，含有鮮奶成份，如勾芡汁般處理。西式的麵撈，即熱熔牛油再放入麵粉炒至金黃，再逐少加入上湯推至有筋性，待接近完成後才放入淡奶調至完全融和。優點是不易回水，非常稠密，並可預先處理，用時取一點，多加點湯變稀，就是忌廉湯了。

每菜一食材

石斑

石斑種類很多，分老鼠斑、紅斑、青斑、油斑、杉斑、芝麻斑、泥斑、瓜子斑、雜斑和黃釘等。

在香港以老鼠斑最珍貴，紅斑次之，能夠出得場面又味美，價錢適中的，非青斑莫屬。

單一料：石斑

材料：

石斑1條約1斤（600克），或其他任何鮮魚
蛋白1隻
雲腿絲少許
芫荽少許

調味料：

淡奶100克
上湯100克
雞粉1茶匙
粟粉2茶匙
薑汁酒1茶匙

做法：

1 魚洗淨，抹乾備用。

2 於鑊中注入油，份量要可以蓋過魚。

3 將油燒至八成滾，放下鮮魚，浸 2 分鐘後，即行收火，再浸約 7-8 分鐘，見魚眼突出，
表示已熟，可取出，隔去油。

4 將調味拌勻煮滾，熄火，下蛋白，加入少許尾油及火腿絲，淋在魚面，放上少許芫荽
即成。

子薑百香果醬肉

成菜特色

色彩鮮艷，外酥內嫩，酸甜微辣適口。

☑ 10分鐘　👤 4~6人

烹調要點

1 油炸是烹法的成敗關鍵，首先油量要充足，炸油要過物料之頂面，火要大，油要熱，如採用一次過的炸法，油溫應在七八成之間。

2 採用炸兩次的方法，則第一次炸時，油溫略低，約六成左右，炸至七成熟為度，第二次複炸時，油溫要高時間要短，能將表面炸酥脆為佳。

每菜一食材

五花肋條肉

肥瘦肉有規則的相間排列，形成三層或五層的肋條肉，多用於紅燒、炆、炸或粉蒸。

其特質是肥瘦相間，肉質因有肥油隔層，肥美幼滑，肉汁很多，味道濃香，又因用途廣泛，無骨，頗受食客和廚師歡迎，加上它的入饌可索性高，許多經典菜式都以它為主角。

主料：混合料，即主輔料份量相約

材料：	醃料：	調味：	
腩肉300克，切塊	醬油1茶匙	水4湯匙	雞粉1茶匙
蜜桃80克，切塊	蛋黃1隻	白醋1湯匙	百香果醬2湯匙
紅椒仔1隻，切粒	鹽¼茶匙	黃砂糖2湯匙	山楂餅4片
青椒60克，切塊	雞粉½茶匙	噲汁1茶匙	生粉1茶匙
酸子薑20克	麻油少許		
生粉60克，上粉用	胡椒粉少許		
	生粉1茶匙		

做法：

1 將腩肉和1茶匙梳打粉、1湯匙糖拌勻，醃約1小時，洗淨，瀝乾。

2 調勻醃料，加入腩肉醃約15分鐘以上。

3 將醃好之豬肉倒進笒箕中，使汁水流出，再撲上乾生粉，放八成滾油中，炸至金黃香脆，盛起瀝油。

4 爆香配料，注入調味待滾，立即熄火，傾下已炸好之肉，快速拌勻，即可奉客。

奇香鱔球

金黃油亮、外酥內嫩、鮮香可口。

烹

☑ 10分鐘　👤 4~6人

烹調要點

1 鱔魚腥味較重，醃漬前應洗淨血污，如要紅燒烹製，宜多放紹酒，而炸焗則宜放玫瑰露。

2 鱔魚多幼骨，所以請魚檔起骨時，小心處理，如這菜有小孩或老人享用，買回來後再小心挑骨一次，否則暗藏幼骨，容易刺損喉嚨。

3 鱔魚買回來，先用點鹽在內外擦洗，會出現一層奶白色的濃潺液，再沖走鱔魚上的潺液，重複數次，還可燒點微沸水，熄火後浸入白鱔魚片刻，徹底去除潺液就可使用。

每菜一食材

白鱔

白鱔簡稱"鰻"或"鰻鱺"，營養價值極高，慢性消耗性疾病，如：肺結核、淋巴結核、慢性潰瘍等疾病，均有較顯著的食療作用，白鱔肉質細嫩，味鮮美，是食用魚類的上品。

單一料：鱔

材料：	醃料：	調味：
白鱔1½斤（900克）	醬油2茶匙	沙律醬3湯匙
炒香芝麻適量	鹽焗雞粉2茶匙	吉士粉1茶匙
生粉適量	玫瑰露酒1茶匙	水4湯匙
	熟油1湯匙	煉奶1茶匙
	生粉1湯匙	檸汁1茶匙

做法：

1 將鱔洗去黏液，去骨，洗淨，在魚肉那面剁格子紋。

2 以醃料拌勻，炸前撲上適量乾生粉。

3 放八成熱油中，炸至金黃盛起。

4 煮滾調味，把鱔球回鍋，快速拌勻上碟，灑上少許炒香芝麻即成。

椒麻魚塊

滷汁濃而不膩，完全包裹魚塊。

熘

⏱ 30分鐘　　👤 6~8人

烹調要點

1. 鯇魚腩的骨比較少，肉質肥美幼嫩，就算煮過火，其肉仍鮮嫩，加上它的味道 甜，頗適合與不同味型的配料或香料提味。

2. 花椒的味道濃郁，特質是香而帶麻辣，十分嗆鼻，但在味道濃郁的菜餚中，就不可缺，本菜譜的花椒份量已足夠，宜減不宜加，否則味道過濃。

3. 這是以濃汁入魚，再以香料回鑊提味的菜餚，即煮即吃，方能品嚐真味。

每菜一食材

老抽

滷水除了香料為主，老抽就是調製黑滷汁的靈魂，它的味道不會太鹹或太濃，主要用作調色之用。市面的品種很多，草菰老抽的色澤暗啞、香味不足，鹹味夠，許多食肆廚房愛用；家庭式的老抽，色澤微金黃，味道適中，鹹味略輕，新加坡的老抽或醬油，色澤深帶點棕紅，滷出來的成品就有美麗色澤，許多高級食肆最愛使用。

單一料：草魚

材料：	滷水汁：	醃魚料：	滷水料：
鯇魚腩2斤（1200克）	醬油5湯匙	醬油2湯匙	八角2粒
粗薑條數片，醃魚用	片糖40克	鹽½茶匙	花椒2茶匙
葱2條，切段，醃魚用	鹽½茶匙	薑汁酒1湯匙	薑2片
花椒粉½茶匙，最後回鑊用	水1量杯	熟油1湯匙	葱1條
薑米1湯匙，最後回鑊用			
蒜米1湯匙，最後回鑊用			
麻油1湯匙			

做法：

1 洗淨魚，切成大塊，加入薑、葱及醃魚料，拌勻醃30分鐘，抹乾，備用。

2 燒熱油，爆香滷水料，加入滷水汁煮沸，以慢火熬成濃汁約½杯，隔去渣後備用。

3 燒紅鑊加油，約八成熱，將魚放下，炸至金黃硬身撈起。

4 將魚放入滷水汁內浸泡至汁略收乾，即可盛起。

5 燒熱1湯匙麻油，加入花椒粉、薑米和蒜米，炒至香味溢出。

6 將魚放下快速炒勻上碟。

參巴茄子

醬紅油亮，椒麻辛香，鹹鮮適口。

燜

☑ 15分鐘　👤 4~6人

烹調要點

1. 當魚塊放入熱油中，由於沒有上粉的原固，魚皮很易會黏在一起，這時只要將魚放在滷汁中，便可泡開，製成品便能完整無瑕。

2. 茄子帶有鹼性，最好先以滾油略炸，這樣不但能保持茄子原有色彩，還可增添滑嫩口感。

每菜一食材

茄子

茄子又名"落蘇"，俗稱"矮瓜"，是一種含漿瓜果，色澤有紫色和白色，入秋後又糯又甜。茄子含有多種維他命、蛋白質和鈣，有散血、止痛、消腫作用，更有防止血管破裂和平血壓攻效。不過，茄子性帶寒，生吃會腹痛肚瀉。

主料：茄子；**輔料**：番茄、洋葱

材料：	香料：	調味料：
茄子500克	尖嘴紅椒70克	番茄醬2湯匙
番茄100克	辣椒仔20克	喼汁1湯匙
洋葱1個	乾葱頭100克	糖、鹽各½茶匙
芫荽1棵	蒜頭50克	雞粉1茶匙
炒香白芝麻適量		清水¼杯
		HP醬1½茶匙

做法：

1 茄子切塊，用滾油炸約1~2分鐘，盛起瀝油。

2 洋蔥切塊、芫荽切碎、番茄切塊。

3 燒油3湯匙，炒香料至香味溢出，加入洋蔥炒透注入調味料煮沸。

4 將茄子回鑊拌勻，加入番茄煮2~3分鐘，加入芫荽拌勻上碟，灑上芝麻即成。

醋熘明蝦球

成菜特色

嫣紅油亮，酸甜適口。

☑ 10分鐘　👥 4~6人

烹調要點

1. 用蝦來做的菜餚，通常都要先泡油，再與其他配料合炒，泡蝦時需用紅鑊猛油，否則蝦便不夠爽脆。

2. 雲耳屬菌類，本身會散發一股獨特的霉味，所以浸發後用少許生粉擦洗，再用清水沖洗乾淨，再飛水，可降低其異味。

3. 蒜芯或青豆仁，如欲保持翠綠，可泡嫩油，才加入與其他材混合快炒，色澤會美麗點。

4. 用生粉撲在蝦後才泡油，確保蝦肉的汁液不易流失，外乾脆而內存肉汁。

每菜一食材

雲耳

性質與黑木耳相同，但形較小而薄，較透明軟滑，易消化，宜用炒、蒸、汆等烹調法。

其特質沒有濃厚味道，但卻有一種獨特香味，加上屬於養生健康妙品，對血脂高的人，有明顯幫助。

主料：蝦；**輔料**：雲耳、青豆

材料：	醃料：	調味：	
中蝦300克	蛋白1茶匙	鎮江香醋1茶匙	山楂餅10克
濕發雲耳80克	鹽¼茶匙	蠔油1湯匙	雞粉1茶匙
蒜芯或青豆仁40克	生粉1茶匙	醬油1茶匙	生粉2茶匙
蒜頭2粒，切片		茄汁2湯匙	清水½杯
乾蔥頭2個，開邊		花椒油1茶匙	
生粉適量			

做法：

1 鮮蝦洗淨去殼、腸，留尾；蒜芯切度。

2 以醃料拌勻，撲上適量乾生粉，泡滾油備用。

3 燒油2湯匙，爆香蒜片、乾葱及蒜芯，下雲耳炒片刻，加入調味煮滾。

4 將蝦回鑊，拌勻上碟。

松子黃魚

鮮艷奪目,外脆內嫩,酸甜適口。

☑ 15分鐘　👤 4~6人

烹調要點

1　要掌握好味汁調製時間,最好與原料熟製的過程同步進行,即一邊炸製原料,一邊調製味汁。"焦熘"菜餚必須趁熱食用,讓食客感受到裹滿熘汁的原料,入口還是那麼香脆可口。

2　現在的黃魚多以人工飼養為主,少有野生,加上魚小肉薄,必須挑選大條肉厚,才容易起魚。

3　醃料用檸檬汁可減去魚腥味道,並留有一股清香新鮮味道。

每菜一食材

黃花魚

是中國特產魚類,色如黃菊,頭部生有石狀枕骨,又名耳石,有食療作用,以枸杞、菊花蒸黃花,能明目、開胃、益氣,但有傷風咳嗽、氣管炎者忌食為佳。

主料:黃花魚;**輔料**:松子、番茄、青豆

材料:	醃料:	調味料:
黃花魚1條,	檸檬汁1茶匙	茄汁½杯
松子1湯匙	熟油1湯匙	糖1½湯匙
番茄粒¼杯	鹽½茶匙	吉士粉1茶匙
青豆仁1湯匙	雞蛋½隻	雞粉½茶匙
生粉適量		麻油少許
		清水¼杯

做法：

1 魚洗淨起骨，用醃料拌勻15分鐘後，上乾生粉備用。

2 燒油至八成滾，放下魚塊炸至金黃盛起，置碟中。

3 煮滾調味，加入番茄粒及青豆推勻後，淋在魚面，最後灑些松子仁即成。

馬拉盞煎黑鯧

色澤金黃，外脆內嫩，甘香辛辣適口。

成菜特色

✅ 20分鐘 👤 4~6人

烹調要點

1 黑鯧全身黑褐色，肉質和食味與白鯧略同，但沒有白鯧般嫩滑，故用來煎炸較佳。

2 "馬拉盞"亦稱峇拉煎，是選用新鮮幼蝦製成，一般地方稱為"蝦膏"。

3 不一定用鯧魚，其他的如河魚、鮫魚等也可以做。

4 煎時要慢火，並小心不要讓餡料溢出，以免容易燒焦。為方便煎製，一般選擇魚體扁平狀為佳，煎時要慢火，並小心不要讓餡料溢出，以免容易燒焦。

每菜一食材

黑鯧

黑鯧，全身黑褐色，肉質和食味，與白鯧略同，但沒有白鯧般嫩滑，一般用來煎炸較佳。

單一料：黑鯧

材料：
黑鯧魚1條(約400克)

調味料：
糖1½茶匙
檸汁1茶匙
李派林辣椒汁1茶匙
鹽½茶匙

醬料：
紅辣椒3隻
馬拉盞20克
蒜頭、乾葱各4粒

做法：

1 將魚劏好洗淨。

2 搗爛全部醬料，加入調味料拌勻，釀入魚肚中。

3 將魚以慢火煎至兩面金黃，即可上碟。

乾煎生曬蠔豉

色澤金黃，甘香可口。

☑ 15分鐘　👤 4~6人

烹調要點

1 買回來的蠔豉，在未烹調前，要放入雪櫃冰格內才不易變壞，若處理不善，容易發霉，色澤暗啞，不宜食用。

2 浸蠔豉的清水不要太多，否則會令其鮮味因浸泡而流失。

3 浸蠔豉前先清洗，再用水浸發，可保留浸水做其他菜式，但不要完全採用，因為蠔豉的沙泥或碎殼會留在水底，可棄掉。

每菜一食材

蠔豉

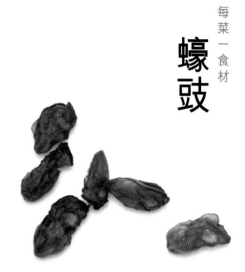

選購生曬蠔豉，以蠔身肥美，飽滿肉滑，鮮明乾爽為佳，如蠔豉呈深啡色，多是舊貨，此類蠔豉必是粉重，容易爆開，入口鬆散，無嚼頭或有糠味，只可作煲湯用料。

單一料：蠔豉

材料：

生曬蠔豉12隻
生粉適量

醃料：

薑汁酒1湯匙
蜜糖2茶匙
檸汁1茶匙
醬油1茶匙
紅辣椒1隻，切粒

做 法：

1 洗淨蠔豉，用過面清水浸過夜，盛起瀝乾。

2 醃料調勻，放入蠔豉，醃30分鐘。

3 將蠔豉兩邊薄薄撲上乾生粉。

4 燒熱適量油，將蠔豉排入鑊中，以慢火煎至兩面金黃即可上碟。

香煎蘿蔔墨魚餅

一面金黃香脆，一面軟糯清鮮。

貼法 ✓ 15分鐘　👤 4~6人

烹調要點

1. 貼法的用油量較煎法為多，但油量只可達到相當於原料厚度的一半，煎時要不斷將鍋中熱油澆潑在原料上。

2. 市面售賣的墨魚膠很多，如果鮮度不足，出現一種異味，就不要用了。

3. 凍肉店有墨魚膠，如果不是即時應用，可放冰格貯藏，待用時才取在冰箱溶解，或是置水喉下沖洗表面，待其自動解凍，或是浸入冷水解凍。

每菜一食材

白蘿蔔

白蘿蔔別名萊菔、土酥。生者味辛、性寒，熟者味甘、溫平，以表皮光滑、結實飽滿而無裂痕為新鮮嫩蘿蔔，不用清洗，以保鮮袋盛着放冰箱，可保存一星期左右。

在食療方面，它有清熱化痰、生津止渴和抗菌作用，汁液可治療、預防膽結石，而所含的糖化酵素、木質素，有抗癌防癌功效。

主料：蘿蔔、墨魚膠；**輔料**：香芹、辣椒

材料：
白蘿蔔1個300克
墨魚膠240克
香芹(切粒)2湯匙
辣椒(切粒)1隻
清水1杯，煮蘿蔔用

調味：
鹽½茶匙
雞粉1茶匙
胡椒粉少許
麻油少許
糯米粉1湯匙

做法：

1 蘿蔔刨粗絲，以1杯清水煮10分鐘，盛起，壓乾，待涼備用。

2 把墨魚膠置盤中，加入蘿蔔絲、香芹、辣椒粒及調味，攪至起膠。

3 在手中塗少許油，將以上用料分成等份，搓成圓球，輕輕按扁。

4 燒熱煎鍋，加入大量油，放下墨魚餅，用慢火一邊將一面煎成金黃香脆，一邊在表面
淋熱油，熟後，切件上碟。

鍋貼香菜蛋包魚

金黃淡雅，鮮香味美，軟嫩可口。

✓ 8~10分鐘　👤 4~6人

烹調要點

1. 最好取一隻約6-7吋之小型煎鍋，這樣才能煎成厚厚的熱香餅狀，用易潔鍋煎則較易扣出。

2. 煎蛋餅時，不要心急，以中火慢烤烘，盡量令表面烘至金黃或乾身，期間需要察看，如因易潔鑊過熱，蛋餅會變焦，可先離火降溫，或下點冷油。

3. 醃料有蛋白，可令提升魚肉味道，更能使其肉質幼滑。

每菜一食材

芫荽

氣味芳香，是一種極之普遍的副食菜蔬，將它加入魚、肉及粥類，能去腥增香及起殺菌作用，此外更能清熱解毒，以馬蹄、紅蘿蔔、竹蔗與連根之芫荽煲成湯水，有去疹毒及清熱治喉痛、喉炎的作用。

主料：魚；輔料：蛋、芫荽

材料：
雞蛋5隻
黃花魚1條
芫荽2棵，切碎
紅椒仔1隻，切粒
薑米1茶匙
生粉適量

醃料：
蛋白1茶匙
味鹽½茶匙
生粉½茶匙
熟油1湯匙

調味：
雞粉½茶匙
鹽½茶匙

做 法：

1 雞蛋打散，加入調味，拌勻備用。

2 洗乾淨魚，起肉切丁，以醃料拌勻，泡油，待凍備用，在頭、尾及魚骨撲上適量乾粉，
炸至金黃排放碟中。

3 將魚肉、芫荽、紅椒粒、薑米等加進蛋液中拌勻。

4 燒熱易潔煎鍋，加入適量油，放下以上材料，以中慢火煎成熱香餅狀。

5 放涼，切成塊狀，放在炸香的魚骨上即成。

蝦子鍋塌豆腐

色澤金黃，軟嫩鮮香可口。

塌

☑ 15分鐘　👥 4~6人

烹調要點

1 塌法的風味質感與煎法大致相同，但與煎相比，出品外部不夠酥脆，內部卻比較軟嫩，這是因為原料經上糊後，在煎製時形成厚膜，故加入調味湯汁後，便能充分吸收，而成外軟內嫩的特色菜餚。

2 豆腐的水份很重，置一旁時會不時滲出水份，撲上生粉前必須抹乾，才容易上粉。

3 用生粉做外皮，可以防水豆腐水份因油炸時滲出炸油中或煎油中，造成濺油而傷身。

每菜一食材

蝦子

將蝦的卵煮熟，曬乾，便成蝦子。蝦子又名"蝦春"。以蝦子烹調的菜餚，如上海名菜"蝦子大烏參"及港式"蝦子麵"等。

主料：豆腐；**輔料**：蝦子

材料：	醃料：	調味：
百福硬豆腐1盒	雞粉1茶匙	雞粉1茶匙
雞蛋黃2個		上湯⅓杯
生粉適量		
蝦子適量		

做法：

1 用布吸乾豆腐的水份，切開6塊；打散雞蛋黃。

2 將豆腐滾上適量乾生粉，再塗上蛋黃液。

3 燒熱適量的油，放入豆腐，以中火煎至兩面金黃色。

4 將調味拌勻，由鑊邊注入，以中慢火收汁（需反轉一次）。

5 上碟後，灑上少許炒香蝦子即成。

香菇彩椒炒斑柳

色澤鮮明，汁液剛包住魚肉，清鮮爽脆。

炒

☑ 10分鐘　👤 4~6人

烹調要點

1 石斑肉連皮切塊，除了令成品有色澤，還可以保護魚肉不易散掉。

2 魚肉放醃料前，先把魚肉抹乾水份，容易入味，時間不要醃太夠。

3 可以加入水果或其他具色澤的配料也可。

每菜一食材

甜椒

甜椒又稱燈籠椒、西椒，色澤有很多，如紫、紅、黃、青、綠、橙等，肉厚味甜，爽脆而不含辣味，可作主料或配料，其外衣比較厚，所以西菜會有燒彩椒，用明火或焗爐燒焦外皮，剝去外皮時椒味濃而清甜，可配三文治、沙律、生吃、炒、焗、釀等烹調方法。

主料：

材料：	調味：	醃料：
石斑肉6兩（240克）	雞粉1茶匙	蛋白1茶匙
冬菇3隻	麻油少量	生粉½茶匙
三色椒各½隻	胡椒粉少量	雞粉½茶匙
蒜頭1粒，切片	水4湯匙	熟油1湯匙
紹酒1茶匙	生粉1茶匙	

做 法 :

1 將石斑肉切條，以醃料拌勻，泡油備用。

2 將冬菇浸軟，切條；把三色椒也切條。

3 燒熱鍋，炒香蒜片，加入冬菇略炒，再加入三色椒炒熟，灒紹興酒。

4 加入調味和石斑條，炒熟，上碟即可。

麻辣雞丁

色彩悦目，醬紅油亮，麻辣適口。

炒 ☑ 25分鐘 👤 6~8人

烹調要點

1 花生要先煮9分鐘，直到水份煮乾後，再取出、完全吸乾。

2 炸花生前，要先炸香料，讓香料的麻辣味融入油中，然後撈出香料，用這些油炸花生，花生便有麻辣味。

3 待花生炸至金黃撈起放涼後，把炸過的香料放回花生中拌勻，再一起存放，花生便會有極佳的麻辣效果。

4 如果用四季豆必須飛水、過冷；蒜心就不用飛水。

每菜一食材

花生

花生，又稱落花生及長生果，是一年生草本植物，開黃花，果實在地下生長成熟，含蛋白質及脂肪，可榨油。中醫認為熟食有補脾潤肺補血之功。

主料：雞肉；輔料：花生、四季豆、辣椒乾

材料：		醃雞料：	調味：
麻辣花生1½兩(60克)	辣椒乾½兩(10克)	蛋黃½隻	醬油½湯匙
蒜心或四季豆2兩(80克)，切粒	蒜茸2茶匙	醬油2茶匙	鎮江香醋½茶匙
	薑蓉1茶匙	蠔油1茶匙	黃砂糖½茶匙
雞腿肉8兩(300克)	生粉適量	生粉½湯匙	生粉½茶匙
		熟油1湯匙	水2湯匙

做法：

1 雞肉切丁，用醃料拌勻，炸前再撲乾生粉。

2 燒油八成熱，放下雞肉，炸至金黃，瀝乾，盛起。

3 燒熱少許油，放下辣椒、蒜茸、薑蓉爆香，傾下雞肉、麻辣花生和蒜心。

4 加調味拌勻上碟。

炒鱔糊

成菜特色

蒜香飄逸，清雅悦目。

⏱ 10分鐘　👥 4~6人

烹調要點

1. 黃鱔的生命力很強，肉質爽脆，用少許鹽擦洗，必須徹底去掉潺液，鱔魚肉才能清爽。

2. 韮黃宜吃半生熟，不應煮得太熟，否則會出水和變黃瘀，賣相和吃味也不好。

3. 傳統的鱔糊，在上桌前放生蒜，淋熱沸油，使之變熟。

每菜一食材

黃鱔

體細長，光滑無鱗，屬合鰓科，與白鱔完全不同類，多用作起肉炒絲或焗飯。

其幼質幼嫩，肉味濃烈，屬於田園河塘的食材。

主料：黃鱔；**輔料**：韮黃

材料：	醃料：	調味：	裝飾：
黃鱔12兩(480克)	醬油1茶匙	醬油3湯匙	胡椒粉適量
薑蓉2茶匙	鹽¼茶匙	雞粉½茶匙	
紹酒2茶匙	胡椒粉少許	生粉1½茶匙	
韮黃2兩(80克)	麻油少許	糖1¼茶匙	
蒜頭3粒，剁碎	生粉1茶匙	麻油少許	
	熟油1湯匙	水⅓杯	

做 法 :

1 黃鱔起骨(可請魚檔代勞),用熱水洗去黏液,洗淨切成1½吋長的粗條。

2 加入醃料拌勻,泡八成滾油備用。

3 燒油2湯匙,爆香薑蓉,灒酒,將鱔條回鑊,以大火炒透。

4 傾下韭黃翻炒至勻,下調味炒透上碟。

5 中央撥開一小洞,把生蒜茸放入。

6 將少許麻油及2湯匙油置鑊中燒滾,淋在蒜茸上,四周灑上適量胡椒粉即成,吃時拌勻。

玉種藍田

清雅悦目，鮮香冶味，爽脆適口。

成菜特色

⏱ 8~10分鐘　👤 4~6人

烹調要點

1 炒菜心和芥蘭，如用慢火，可以不用加水炒，當燒紅鑊後，放油時跟着放菜炒勻。如用大火則要放下少許清水，以免菜裏的水份來不及釋放而煮焦。

2 芥蘭與薑汁酒的味道很配，可以多下點無妨。

3 田雞的肉質幼細快熟，質感如吃雞肉，肉纖維幼嫩，所以利用先泡油，後快炒的烹調方法，最能突顯這菜的特色。

每菜一食材

芥蘭

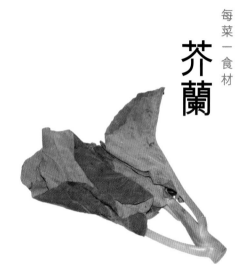

質感爽脆，菜葉表面含有一層銀白灰質，略帶苦澀味道，荷塘芥蘭宜吃梗柄為主，3吋芥蘭薹則以焯、快炒為主，至於一般芥蘭就要梗葉均吃。靚的芥蘭，葉少梗粗，硬挺而外衣粗纖維薄，菜花含苞未開，而菜梗有天然爆口，就清脆鮮甜。

主料：田雞、芥蘭；輔料：鹹魚

材料：	醃料：	調味料：
田雞10兩（400克）	蠔油1茶匙	蠔油2茶匙
芥蘭心6兩（240克）	醬油2茶匙	醬油1茶匙
鹹魚1兩（40克）	雞粉½茶匙	雞粉1茶匙
薑花4片	麻油少許	麻油少許
甘筍花6片	胡椒粉少許	胡椒粉少許
薑汁1湯匙	生粉1湯匙	生粉1茶匙
雀巢1個		清水4湯匙

做 法：

1 田雞洗淨、斬件，用醃料拌勻，15分鐘後泡油備用。

2 將芥蘭心切成1½吋長，在頂部切十字，飛水，過冷，備用。

3 鹹魚蒸熟，切小粒。

4 燒熱油，爆薑花，傾下田雞、芥蘭、鹹魚等炒透，下調味拌勻，以雀巢盛載上桌。

豆芽鬆

清爽鮮嫩，成菜乾爽，主料和輔料完全融合。

成菜特色

☑ 15分鐘　👤 4~6人

烹調要點

1 凡炒豆製食品，必要以白鑊先將豆芽頭部烘乾至發出豆香為佳，這除了可以去豆腥味外，還可增添香氣。

2 炒大豆芽後盛起，待片刻讓其水份排出，棄去不要，再配合其他材料，方能做出菜的精髓。

3 這菜屬粗料精做，講求刀章幼細，並以快炒方法成菜。

4 成菜關鍵以火猛快炒，鑊氣足，火候要拿捏得宜，否則容易出水，失去特色。

每菜一食材

大豆芽

大豆芽是大豆(黃豆)經加工發芽而成，以豆瓣完整而不爛、無豆皮、體短肥壯、質脆為優品。豆芽的蛋白質是一般蔬菜的10倍，還有人體必需的磷、鈣、鐵、鉀等元素，維他命 B_1、B_2 和維他命 C 的含量也高。中醫認為它有清熱、除濕的作用，而據外國的一些報導，大豆芽中含有一種預防癌病的酶，是相當理想的食品。

主料：大豆芽；**輔料**：豬肉

材料：	醃料：	芡料：
大豆芽1斤(約640克)	醬油、蠔油、油各1茶匙	蠔油1湯匙
攪碎豬肉6兩(約240克)	生粉½茶匙	醬油、生粉各1茶匙
香芹1棵，切粒	糖¼茶匙	糖、雞粉各½茶匙
蒜茸1茶匙	胡椒粉少許	麻油、胡椒粉各少許
		水4湯匙

做法：

1 豬肉加醃料拌勻，醃片刻，備用。

2 摘去大豆芽尾端，將梗分半，豆芽頭剁碎。

3 用鑊先烘乾豆芽頭，然後加入1湯匙油，炒香之後，再加進豆芽梗，炒至七成熟盛起。

4 燒熱鑊，爆香蒜茸，加入豬肉炒透，隨即倒入豆芽、芹菜粒及芡料，加芡兜勻上碟。

馬友鹹魚炒雜菜

鹹鮮適口，清爽淡雅。

生炒

☑ 8~10分鐘 👤 4~6人

烹調要點

1 炒雜菜沒有規定用甚麼蔬菜，只要你的冰箱內有多出的蔬菜便可隨意使用，不要浪費。

2 雜菜可應用任何一種，但要懂得其菜質和需要烹調的時間，按需要而分序炒，否則不是變黃或變黑，就是未熟或過熟，沒有誘人的顏色。

3 蔬菜是瘦物，所以要色澤美麗就多下點油炒。

4 鹹魚是提味的配菜，任何品種都可以，但份量則按其鹹度而加減。

每菜一食材

鮮蘆筍

鮮蘆筍含有大量的維生素和纖維，處理烹調方便，味道清香爽甜，由於多是進口菜，故價錢普遍略高。

主料：混合料，即主輔料份量相約

材料：

馬友鹹魚40克	蜜豆80克
蘆筍160克	黃椒80克
銀芽80克	蒜茸1茶匙
茄子80克	薑茸1茶匙
竹筍80克	

調味：

雞粉1茶匙
生粉2茶匙
水2湯匙

做法：

1 鹹魚切粒，以醃料拌勻。

2 將所有蔬菜洗淨，切好。蘆筍削去尾部老硬外皮。茄子要泡滾油，備用。

3 燒油2湯匙，爆香薑茸和蒜茸，加入鹹魚炒片刻至熟。

4 將以上所有蔬菜（除銀芽外）加入，以大火炒熟，最後加入銀芽及已泡油之茄子，翻炒約20秒。

5 加入調味拌勻上碟。

金銀芽炒滑肉絲

成菜特色

淡雅悦目，肉清菜脆，鮮香適口。

☑ 8~10分鐘 👤 4~6人

烹調要點

1 肉眼洗淨，片成厚約0.3厘米之大片，再斜紋切成長絲，上漿醃好後，放入5-6成油溫中，以筷子撥開，防止黏着，一見肉絲變色，即熄火盛起，瀝油備用。

2 勾薄芡或琉璃芡，份量不宜多，應以緊包裹物料，完全掛在菜餚的份量便可。

3 用少量蘇打粉醃肉，可令肉質鬆軟，但不要加入太多，否則完全破壞肉纖維，並會產生一點苦澀味道。

每菜一食材

肉眼

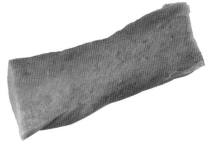

脊下之肉名"肉眼"，又名"里肌"，無筋，食時不韌，肉熟後色白，宜作煎炒之用。

其特質是肉纖維組中帶幼，質感結實，容易切成幼絲。

主料：銀芽、肉絲；輔料：韮黃、冬筍

材料：	醃料 A：	醃料 B：	調味：
肉眼扒4兩（160克）	蘇打食粉½茶匙	蛋白1茶匙	雞粉1茶匙
冬筍1½兩（60克）	清水2湯匙	味粉¼茶匙	鹽¼茶匙
銀芽8兩（300克）		生粉½茶匙	胡椒粉少許
韮黃½兩（20克）		清水2茶匙	清水4湯匙
蒜1粒，切片		熟油1湯匙	糖½茶匙
甘筍絲2湯匙			麻油少許
			生粉1茶匙

做法：

1 肉眼切絲，以醃料 A 拌勻，20 分鐘後，沖洗乾淨，瀝清水份，再以醃料 B 拌勻，泡嫩油備用。

2 冬筍切絲，飛水過冷瀝乾，韭黃切度。

3 燒紅鑊落油 2 湯匙，爆香蒜片，先下筍絲炒片刻，立即傾下銀芽快炒斷生，即可將肉絲回鑊，下韭黃及調味，拌勻上碟。

鮮蛤蜊炒牛奶

成菜特色

潔白素雅、滑嫩味鮮。

⏱ 10~15分鐘　👥 4~6人

烹調要點

1 在炒奶的過程中，必須由始至終以中溫火候順方向炒勻，這樣才不會把鮮奶料炒散或炒焦。

2 炒鮮奶利用火力調控，只把其質由液態變凝固，不能有焦黑或變焦黃，方是上品。所以必須有耐性和懂得掌控火力，太猛時要離火或熄火，太慢時又要適量加大火力。

3 蜆(蛤蜊)烹調前，先灼2分鐘，待蜆殼開口，起出蜆肉，確保肉粒新鮮沒有變壞，較為安全。

每菜一食材

蜆

蜆棲息在鹹淡水交界之河江入海口附近，埋於沙質或泥質內，殼色黃者為肉質肥美，稱為黃沙大蜆。蜆肉有清熱解毒，利濕去黃之功效。

主料：蛋白、鮮奶；**輔料**：蛤蜊、松子仁

材料：

鮮奶200克
蜆肉80克
金華火腿20克
雞蛋白150克
松子仁25克
豬油2湯匙，分兩次下，炒奶用
油2湯匙，分兩次下，炒奶用

奶漿：

鮮奶50克
生粉20克

調味：

味粉¼茶匙

做法：

1 奶漿調勻、蜆肉洗淨飛水瀝乾、金腿切小粒、松子仁炸脆。

2 蛋白置盤中，加入調味拌透。

3 鮮奶以中火煮至大熱，但不能煮沸，離火待凍，與奶漿、蜆肉、金腿等加進蛋白裏，
一同攪拌均勻，即成炒奶料。

4 以中火燒熱易潔煎鍋，加油1湯匙及豬油1湯匙，燒至六成熱，即將炒鮮奶料傾下，
用勺不停地順一方向炒，邊炒邊向上翻動。

5 再將餘下之1湯匙油加入，當鮮奶料炒成半凝固時，加入松子仁，淋下剩餘的1湯匙
豬油，繼續炒至全凝固，即以深碟盛起，在碟內堆成山形便成。

滑蛋炒牛肉

炒蛋金黃而保有少量蛋汁，肉片鮮嫩。

成菜特色

⏱ 8~10分鐘　👤 4~6人

烹調要點

1 牛肉炒前必須要泡嫩油，油溫達至六成時，即將牛肉落鍋急炒，見肉色變淡，可撈起瀝油。

2 牛肉要切橫紋，如以直紋切割，肉纖維很粗韌，難於咀嚼。

3 炒牛肉宜用牛冧肉、牛柳邊或肥牛肉，爽嫩幼滑。

每菜一食材

雞蛋

一隻蛋中含有三大物質，就是蛋殼、蛋白和蛋黃。以下是分析各樣之重量：

全蛋平均重量58克，蛋白佔全蛋重量之58.5%，蛋殼佔10.5%，蛋黃則佔31.0%。

主料：混合料，即主輔料份量相約

材料：	醃料 A：	調味：	醃料 B：
牛冧肉3兩(120克)	食用梳打粉½茶匙	雞粉½茶匙	醬油1茶匙
雞蛋5隻	清水¼杯	薑汁1湯匙	糖¼茶匙
葱粒2湯匙		鹽½茶匙	雞粉½茶匙
		白酒1湯匙	生粉½茶匙
			蛋黃½隻
			清水1茶匙
			熟油1湯匙

做法：

1 牛肉順橫紋切薄片，以醃料Ａ拌勻，醃1小時，洗淨，瀝後，再以醃料Ｂ拌勻備用。

2 雞蛋打散，加入調味及葱粒拌勻。

3 牛肉泡嫩油，盛起瀝油，倒進蛋液中拌勻。

4 燒熱煎鍋，傾下約4湯匙油，將全材料傾下，以大火急炒至蛋液半凝固，便可上碟奉客。

醬爆春花卷

色彩繽紛，醬香飄逸，微辣適口。

爆

☑ 10~15分鐘 👤 4~6人

烹調西女點

1. 爆的特點是"三旺三熱"，即灼燙時要大火滾水，炸時要大火滾油，回鍋時要紅鑊熱油，這三個步驟都要快，才能做出水準。

2. 豬柳肉又稱無骨豬排肉，肉質爽嫩，沒有太多脂肪，很適合爆炒方法處理。如果家中有老人，可改用柳梅肉，幼滑細緻，沒有脂肪，肉纖維幼嫩腍滑，不需要用力咀嚼也可以，但肉味略清淡。

每菜一食材

馬蹄

馬蹄(荸薺)可生吃，也可熟吃，將其加工製成馬蹄粉，可供烹調上多種用途，以廣西桂林或廣州泮塘出產的為最上乘。馬蹄有清熱，利尿，去濕的功效，若眼球赤紅，口滾鼻熱，喉嚨作痛，飲用鮮馬蹄汁，立即有改善。

主料：豬柳；輔料：筍、馬蹄、合桃

材料：	調味：	餡料：	醃料 A：
豬柳肉160克	醬油1茶匙	合桃肉40克，研碎	蘇打粉½茶匙
筍肉60克，切片	蠔油1茶匙	炒香芝麻1湯匙	糖2茶匙
馬蹄肉4個	糖½茶匙		水4湯匙
蜜豆80克	雞粉1茶匙	**醬料：**	
草菇80克，開邊	生粉1茶匙	豆板醬2茶匙	**醃料 B：**
甘筍花6片	麻油少許	蒜茸1茶匙	蛋白1湯匙
薑花4片	胡椒粉少許		鹽、糖¼茶匙
尖嘴紅椒1隻，切塊	水5湯匙		生粉1½湯匙
			水2茶匙
			熟油1茶匙

做法：

1　將豬柳肉切成 12 塊，以醃料 A 拌勻，40 分鐘後以清水沖淨，瀝乾。

2　將豬柳用醃料 B 拌勻，15 分鐘後備用。

3　將少許生粉鋪在碟上，放上肉片，加些餡料，包成卷狀。

4　將筍肉、蜜豆、草菇，分別飛水過冷瀝乾。

5　在肉卷上撲乾生粉，放入八成滾油中，炸成金黃脆身，盛起。

6　燒油爆香薑花、醬料，加進所有配菜炒透，潑酒，下調味，待滾，即加入肉卷，快速
　　拌勻上碟。

麻辣芥醬田雞

成菜特色

碧綠脆嫩，醬香飄逸，鮮味可口。

爆

🕙 10~15分鐘 👤 4~6人

烹調要點

1. 田雞容易有寄生蟲，故必須煮至全熟才可食用。

2. 專業廚師愛把食材泡油，保持色澤和鎖住食材水份，但潮流講健康，或是家庭式做法不愛油炸食物和份量少，可改用飛水程序、以少量油爆料或微油量煎食材，效果相若。

3. 任何梗柄粗的蔬菜也可做這菜，只要其水份少，質感爽脆均可。

每菜一食材

田雞

田雞為水產兩棲動物，生於魚塘、池沼及一般沒有水有草的野外。採購田雞要挑選肚瘦腿肥的，每斤約可得肉11-12兩，若肚大而腿肥大的，每斤約可得肉7兩左右。

主料：田雞；輔料：芥蘭

材料：	醃料：	調味：	
田雞2隻，冬菇3隻	薑汁酒1茶匙	芝麻醬1茶匙	花椒油1茶匙
芥蘭薳160克	醬油1茶匙	芥辣醬1茶匙	麻油少許
甘筍花8片	蠔油1茶匙	胡椒粉少許	雞粉1茶匙
蒜頭2粒，切片	糖½茶匙	糖½茶匙	生粉1茶匙
薑花4片	熟油1湯匙	水4湯匙	
薑汁酒1湯匙	生粉2茶匙		
糖½茶匙，調勻			

做 法：

1 田雞洗淨斬件，以醃料拌勻，20分鐘後泡油至八成熟。

2 冬菇浸透切片，芥蘭薳切段。

3 燒紅鑊加油2湯匙，爆香薑花，潷薑汁酒，傾下芥蘭薳、冬菇和甘筍花等炒片刻，加水2湯匙，盛起瀝去水份。

4 燒熱適量油，爆香蒜片，將田雞回鑊，以大火炒片刻，即可加入調味及芥蘭，拌勻上碟。

油爆金球

色澤悦目，脆嫩爽滑，鮮鹹適口。

爆

☑ 10~15分鐘 👤 4~6人

烹調要點

1 在爆的過程中，火侯和時間必須掌握得恰到好處，才能使三種不同的物料獲得理想效果。

2 雞翼起骨，只要把兩端的骨尖去掉，就很容易脫骨，並且把雞肉翻出。

3 鴨腎的肉厚，不易炒熟，所以中菜會把它剞花，除了增加賣相，還可以省減烹調時間。

每菜一食材

鴨腎

呈淡紅色，表面有一層薄膜，柔潤有光澤，富彈性。曬乾的鴨腎俗稱陳腎，用作煲湯，冶味健胃。

主料：混合料，即主輔料份量相約

材料：	醃料：	醃蝦料：	調味料：
雞中翼6隻	醬油1茶匙	蛋白1茶匙	蠔油1茶匙
鴨腎3個	蠔油1茶匙	鹽⅛茶匙	醬油1茶匙
中蝦6隻	雞粉½茶匙	生粉½茶匙	雞粉1茶匙
西芹80克	生粉1茶匙	熟油1湯匙	麻油少許
茭白筍80克	熟油1湯匙		胡椒粉少許
甘筍花6片	水1茶匙		生粉1茶匙
蒜肉2粒，切片	玫瑰露1茶匙		水4湯匙

做法：

1 雞翼去骨反轉，鴨腎起花，以醃料拌勻，泡油備用。

2 蝦去殼起雙飛，以醃蝦料拌勻，泡滾油備用。

3 西芹切塊，茭白切角，飛水，過冷，瀝乾。

4 燒油2湯匙，爆香蒜片，傾下西芹、茭白、甘筍花炒勻。

5 將所有材料回鑊，以大火炒熟，下調味拌勻上碟。

爆炒香草鮮鴨片

油潤包汁，香鮮脆嫩，冶味適口。

爆

✓ 10~15分鐘 👤 4~6人

烹調要點

1 以大火高溫熱油，快速操作，爆炒時間不能超過20秒，成菜要內香外熱，質感脆嫩為佳。

2 鮮鴨胸可在凍肉店購買，肉質腍軟。如果對食物有要求，就可以買一隻冰鮮鴨取胸肉。

3 不用鮮鴨胸，可改用煙鴨胸，但就要刪去醃料，否則味道太濃烈。

每菜一食材

九層塔

又稱金不換及羅勒，有野生或培植，分布於亞洲熱帶，可作菜餚調味，能去腥增香，有去濕消食，行氣活血及解毒功效。

主料：鴨胸肉；**輔料**：西芹、紅椒、薑

材料：	醃料：	調味：
鴨胸肉1塊（160克）	醬油2茶匙	醬油1茶匙
西芹80克	香茅1支，略拍	糖½茶匙
紅椒40克	蜜糖1茶匙	油少許
黃椒40克	生粉½茶匙	百里香草粉⅛茶匙
九層塔2棵，取葉	熟油1茶匙	蠔油1茶匙
蒜肉1粒，切片		雞粉1茶匙
		生粉1茶匙

做法 :

1 鴨胸肉洗淨以醃料塗勻，30分鐘後，整塊泡油至金黃，盛起待凍切厚片。

2 西芹、紅黃椒切條。

3 燒2湯匙油，爆香蒜片，傾下西芹、紅黃椒及鴨片，九層塔等，以大火爆片刻，即下
調味拌勻上碟。

麻辣蒜香雞

醬紅油亮，麻辣鮮香適口。

爆

☑ 10~15分鐘 👤 4~6人

烹調要點

1 爆菜應注意的問題是火力，每一環節都要以強烈的剛性火侯操作，但一般家庭式的爐具都不能達到以上效果，惟有將火力開至最猛，以快速手法完成。

2 可用雞胸肉取代雞腿肉。

3 為了健康用去皮雞肉，但味道略清淡和缺了脂肪香味。偶而為之，可連雞皮做這菜，味道更好。

4 鮮雞肉的味道比急凍雞肉好，但價錢略貴。

每菜一食材

辣椒

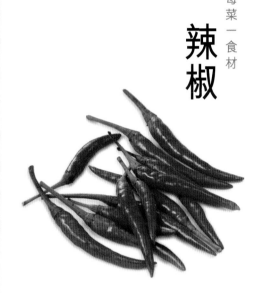

別名"番仔薑"、"紅辣椒"、"番椒"。應選購表面飽滿，光滑，蒂頭的部分鮮綠為最佳。保健功效是開胃暖胃，促進血液循環，防癌。

主料：雞肉；**輔料**：辣椒、蒜芯

材料：	醃料：	調味：	
雞腿肉200克	蛋黃½個	雞粉1茶匙	黃砂糖2茶匙
蒜芯160克	醬油1湯匙	鎮江香醋2茶匙	醬油1湯匙
尖嘴紅椒40克	芫荽粉½茶匙	麻油¼茶匙	花椒油½茶匙
辣椒乾3隻	鹽¼茶匙	生粉1茶匙	水4湯匙
蒜肉6粒	糖¼茶匙		
薑茸1茶匙	生粉2茶匙		
生粉適量	熟油1茶匙		

做法：

1 蒜芯切度；紅椒切塊；辣椒乾切段。

2 雞肉洗淨切塊，以醃料拌勻，炸前撲上適量生粉。

3 燒油至八成熱，放下雞塊，炸至金黃，盛起瀝油。

4 燒油2湯匙，爆香蒜肉、薑茸、辣椒乾，傾下紅椒和蒜芯，炒約20秒。

5 將雞塊回鍋，以大火快速炒透，下調味拌勻，收汁上碟。

拔絲紫心薯

成菜特色

晶瑩明亮，鬆脆味香。

⏱ 10分鐘　👤 4~6人

烹調要點

1 糖與水的比例必須正確，一般用於拔絲的糖液不宜太濃，比例為糖50克，水約為18克左右。

2 熬糖溫的火力不能過大，要以中慢火為度，油溫要在四成熱左右。

3 熬製時間要短，尤其在拔絲時，只在一瞬間，主要是靠視力。

每菜一食材

紫薯

紫色的番薯，食用部份為地下莖，以澱粉為主要部份。番薯含鉀與鈉，在體內有抵抗的作用，雖然番薯味甜，但食番薯等於添加鹽。

主料：紫薯；輔料：糖

材料：
紫心番薯200克
白糖100克
清水36克

脆漿料：
炸粉160克
糯米粉1湯匙
吉士粉10克
清水適量
油30克

做 法 :

1 將紫薯去皮,切丁方大粒。

2 脆漿料置盤中,加入適量清水,開成麵糊,最後加入生油拌勻備用。

3 將紫薯上麵糊,投入八成熱油中,炸至金黃,盛起。

4 白糖、清水置易潔鍋中,以中慢火煮滾,用手勺翻攪糖液,使其受熱均勻。

5 見糖溶液由淺黃大泡,轉為深黃色小泡時,以筷將一滴糖液滴入冰水中,如成半凝固 狀立即離火,然後將炸脆的紫薯放入,快速拌勻。

6 預先將碟塗油,將沾上糖漿的主料傾落碟中上桌,以筷子夾出主料,拔出糖絲,再投 入冰水裏一泡,便成香脆可口的甜品。

反沙芋

潔白如霜，香甜鬆脆，口感細緻。

☑ 10~15分鐘 👤 4~6人

烹調要點

1 掛霜的加工處理方法有多種，除掛糊油炸外，還可乾炸，清炸或先蒸熟後炸，掛霜糖溶液比拔絲要濃一些，才會泛起白霜，熬製時間較短，不能變色。

2 煮糖水的溫度要控制得宜，糖溫約在110~120℃，把糖水滴在冷水中，慢慢沉下，結糖珠，但不能太硬便好。

3 油炸後的番薯而趁熱放在糖水中，不斷快炒，才能做到效果。

每菜一食材

芋頭

小的稱"芋艿"，常見的有荔甫芋、白芽芋和紅芽芋，皮內有鹼性黏液，因鹼性較強，雖沒毒性，卻容易引起皮膚過敏，治療方法是在敏感處放近火，略烘，再以熱水泡洗便可。

主料：香芋

材料：

芋頭200克

乾葱6粒

糖90克

水30克

油適量，要蓋過料頂面

做法：

1 把乾葱切開一半，以適量油炸成金黃香脆盛起。

2 把芋頭去皮切條，投入炸過乾葱之油中，炸成金黃硬身盛起。

3 糖、水置易潔鍋中，以中慢火熬至糖漿起泡，見晶粒聚集變大，便可將炸好之芋條傾下，快速拌勻，立即熄火降溫。

4 待冷卻後，芋條便會自然散開，裝盤上桌。

高麗豆沙香蕉

金黃鬆軟、香甜有彈性。

高麗

☑ 10分鐘　👤 8人

烹調要點

1 蛋白打好後要立即用，如果放的時間過久，蛋白便會瀉身變回蛋清不能用。所以應在慢火燒油時，才開始打。

2 豆沙不可做粗粒，如果蛋球要做成雪糕球般大小，豆沙只做成約波子一般即可。原因是蛋白較輕，豆沙較重，放入蛋白中，可能會從蛋白的另一面掉出來。

3 油溫不宜太熱，應以浸炸方式，即6-7成油溫之暖油即可。

4 炸好之蛋球呈淡黃色，顏色不會太深。

5 灑上糖霜，趁熱食用，涼後會有腥味。

每菜一食材

蛋

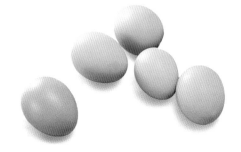

蛋分雞蛋、鴨蛋、鴿蛋、鵝蛋、鵪鶉蛋等。買蛋時，用手執着蛋的兩端，對光照着，是透明的便是新鮮蛋，如混濁的是舊蛋。又可用手執着蛋搖動，有聲音者為壞蛋。

新鮮蛋殼清潔，殼面粗糙，光澤亦不明顯，蛋黃清晰，蛋白稠滑。

主料：蛋白；**輔料**：豆沙

材料：

蛋白5隻（約150克）

麵粉20克、粟粉40克篩勻

豆沙40克

香蕉½隻，切粒

糖霜適量

做法：

1 將豆沙分成8等份，每份包入一粒香蕉，搓圓備用。

2 將蛋白打成厚忌廉狀，加入篩勻之麵粉和粟粉，拌勻成蛋白糊。

3 用雪糕勺，先舀一半蛋白糊，放入一粒豆沙香蕉。

4 再舀滿一勺蛋白糊，放6-7成溫油中，以慢火浸炸至全熟，呈微黃色，盛起，瀝油，灑上糖霜，趁熱烹用。

Indonesian Deep-fried Chicken Wings 印尼香炸雞翼 (p14)

Main ingredient: Chicken wing

☑ 20 mins 📶 4-5

Ingredients:
5 chicken wings

Marinade Ingredients:
2pcs turmeric, 2pcs kemiri, 2pcs shallot, 2pcs garlic, ¼ cup coconut milk

Seasonings:
¼ tsp pepper, 1½ tsp salt, ½ tsp each of parsley powder and lemon grass powder, 1 tbsp brown sugar, 1 tsp chicken powder, ½ tsp Lea & Perrins worcestershire sauce

Procedures:
1. Wash chicken wings and pat dry.
2. Mix all marinade ingredients in blender to make marinating sauce.
3. Place chicken wings in platter. Add in marinating sauce and seasonings. Mix well and marinate for 1 hour.
4. Heat some oil in wok. Deep fry chicken wings until they turn golden brown and are slightly dry.

Deep-fried Fish Cake 炸魚餅 (p16)

Main ingredient: Fish; The minor: String bean

☑ 15 mins 📶 4-6

Ingredients:
240g fish fillet, 20g string bean, ½ egg, 1 tsp Kaffin lime leaves (Thai lemon leaves)

Seasonings:
1 tbsp chili paste, ¼ tsp each of soy sauce, salt, sugar and chicken powder, ½ tbsp cornstarch, ½ tsp oil, Dash of pepper and sesame oil

Dipping sauce:
2 tbsp each of fish sauce and palm (coconut) sugar, 3 tbsp tamarind juice, 1 tsp each of minced garlic, chopped red chili and chopped onion, A little toasted peanut

Procedures:
1. Mince fish fillet, dice string beans, and chop Kaffin lime leaves. Mix these ingredients with egg in large bowl, and then add in seasonings. Stir until the mixture becomes sticky.
2. Slightly wet your hands with oil, and shape the mixture into 10 little fish cakes.
3. Put fish sauce, palm sugar and tamarind juice into saucepan and boil for a while. When cool, stir in remaining ingredients to make dipping sauce.
4. Heat some oil in wok, and then deep fry fish cakes over medium heat until they turn golden brown. Serve with dipping sauce.

Bean Curd and Nori Wonderland 紫竹仙蹤 (p18)

Main ingredient: Bean curd sheet; The minor: Nori

☑ 40 mins 📶 4-6

Ingredients:
600g bean curd sheet (about 10pcs), 10pcs nori, 160g ginger, 6 cups water, Baking paper

Seasonings:
2 tbsp salt, 2 tsp Aji-no-moto

Procedures:
1. Wash ginger and grind it into ginger juice.
2. Pour 6 cups of water into pot. Add in seasonings and ginger juice to make stock.
3. Fold bean curd sheets into triangular shape.
4. Spread boiling stock onto folded bean curd sheets. Drain off excess stock.
5. Lay down bean curd sheets. Add a piece of nori at the centre of bean curd sheet, and then fold into rectangular shape.
6. Wrap it up with baking paper. Put into steamer and steam over high heat for 30 minutes. Take it out, let cool, and then refrigerate.
7. Deep fry bean curd sheet with nori to golden brown before serving.

Spicy Prawns with Garlic 香蒜椒鹽明蝦 (p20)

Main ingredient: Prawn

Ingredients:
600g large prawn, 1pc red chili (diced), 2 tsp minced garlic, 1 tsp Sichuan peppercorn (shallow fry in hot, dry wok), 2 tsp salt (shallow fry in hot, dry wok), Some cornstarch

Marinade Ingredients:
½ tsp Aji-no-moto, A pinch of pepper, A little sesame oil

Procedures:
1. Cut away prawns' legs and remove their intestine. Wash and dry.
2. Saute Sichuan peppercorn in hot, dry wok. Add in salt, and saute over low heat until the colour of mixture changes.
3. Sift fried Sichuan peppercorn and salt. Add into marinade ingredients and mix with prawns. Marinate for 15 minutes.
4. Coat prawns with cornstarch.
5. Heat some oil in wok. Deep fry prawns over high heat for 2 minutes. Drain and set aside.
6. Heat a little oil. Saute garlic and diced red chili. Put prawns back to wok. Stir fry and serve.

☑ 10 mins ▣ 4-6

Spare Ribs with Tangerine Peel 果香陳皮骨 (p22)

Main ingredient: Spare rib

☑ 20 mins ▣ 4-6

Ingredients:
300g spare rib, ½pc dried tangerine peel (soaked), 2pcs preserved plum (cored), 1pc egg yolk, Some cornstarch

Marinade Ingredients:
1 tbsp oyster sauce, 1 tbsp soy sauce, ½ tsp chicken powder, 1 tsp sugar, 1 tsp rose essence wine

Procedures:
1. Chop spare ribs into pieces. Soak dried tangerine peel and preserved plum, and chop finely.
2. Put all of the above ingredients into bowl. Add in marinade ingredients and egg yolk, and mix well.
3. Coat with cornstarch for deep-frying.
4. Heat up some oil. Deep fry ribs until golden brown and cooked.

Crispy Duck 香酥鴨 (p24)

Main ingredient: Duck

☑ 4 hrs ▣ 6-8

Ingredients:
1 duck, 40g dried shallot, Some cornstarch

Marinade ingredients (stir-fried):
2 tbsp peppercorn, 60g fine salt

Seasonings:
¼ cup Shaoxing wine, 2pcs spring onion, 40g ginger (sliced), 1 tbsp fennel, 2pcs dahurian angelica, 1pc dried tangerine peel (soaked)

Procedures:
1. Remove duck's lung. Wash it thoroughly, and then strain.
2. Rub both interior and exterior of the duck with stir-fried peppercorn salt, and marinate for half a day.
3. Take out the duck, wash it thoroughly, and place on tray. Add in seasonings, and then steam in steamer over high heat until cooked.
4. Take out the duck, and pour out the sauce from its body. Discard spring onion, ginger, dahurian angelica, etc., leave the duck cool, and then coat with some cornstarch.
5. Heat a large amount of oil. Add in dried shallot slices and deep fry over low heat until golden brown, and then take out shallot slices. (Shallot can make the oil aromatic.)
6. Deep fry cooled duck in hot oil until golden brown. Take it out, strain out excess oil, and leave it cool before chopping up. Line the pieces orderly in a duck shape on plate.

River Bean Curd 河仙豆腐 (p26)

Main Ingredients: Bean curd, water chestnut; The minor: Dried Chinese mushroom, Chinese sausage, egg white

☑ 20-25 mins ▣ 4-6

Ingredients:
300g bean curd, 200g water chestnut (peeled), 2pcs Chinese mushroom, 20g Chinese sausage, 2pcs egg white, 1pc green onion (chopped), ½ cup cornstarch, Dash of spicy salt

Seasonings:
1 tsp each of salt and chicken powder, ½ tsp sugar, Dash of sesame oil and pepper

Procedures:
1. Wash water chestnut, and grind into puree.
2. Dice Chinese mushroom and sausage.

3. Combine them with egg white, bean curd, green onion, cornstarch and seasonings. Mix well to make paste.

4. Pour the mixture into platter lined with baking paper, and spread evenly by hands.

5. Steam bean curd mixture over high heat for 15 minutes, remove and let cool. Cut into wedges, and then slightly coat with cornstarch.

6. Heat oil until very hot. Deep fry bean curds until they turn golden brown. Remove and drain well. Sprinkle a little spicy salt on top before serving.

Deep-fried Stuffed Crab Pincers 百花釀蟹箝 (p28)

Main ingredient: Shrimp; The minor: Crab pincer

☑ 15 mins   🍴 4-6

Ingredients:
8pcs crab pincer, 300g shrimp meat, Some cornstarch

Seasonings:
2 tsp cornstarch, ½ tsp salt, 1 tbsp egg white, ¼ tsp sesame oil, A pinch of pepper powder

Procedures:
1. Wash shrimp meat. Remove intestine. Pat dry, and mix in mixer.
2. Add in seasonings. Mix to form shrimp paste, and refrigerate.
3. Stuff shrimp paste onto crab pincers.
4. Coat with some cornstarch, and then deep fry until golden brown.

Deep-fried Guilin Beef Balls 酥炸桂林牛肉丸 (p30)

Main ingredient: Beef; The minor: Water chestnut, fatty pork

☑ 10 mins   🍴 4-6

Ingredients:
240g beef, 80g fatty pork, 2pcs water chestnut flesh (finely chopped), 1pc chili (remove seed and finely diced), Some cornstarch

Egg white paste:
1pc egg white, 2 tbsp cornstarch

Seasonings:
2 tsp oyster sauce, 1 tsp soy sauce, 1 tsp sugar, 1 tbsp cornstarch, 1 tbsp egg white, A little sesame oil, A pinch of pepper

Procedures:
1. Chop beef and fatty pork finely; and put into bowl.
2. Add in water chestnut; chili and seasonings, and mix thoroughly until sticky.
3. Knead into meat balls, coat with egg white paste, and then with cornstarch.
4. Deep fry in hot oil over medium heat until golden brown. Serve with dipping sauce.

Mango Waffle Roll 香芒威化卷 (p32)

Main ingredient: Mango

☑ 25 mins   🍴 4-6

Ingredients:
1pc mango (cut into strips), 1pc banana (cut into strips), 2 tbsp salad sauce, Some waffle paper

Crispy coating paste:
½ cup deep-fried powder, 1 tbsp custard powder, ½ cup cold water, 3 tbsp oil

Egg white mixture:
1 egg white, 2 tbsp cornstarch

Procedures:
1. Peel and pit mango, then cut it into 3 x ½ x ½ inch pieces.
2. Peel banana, cut it into 3 x ½ x ½ inch pieces.
3. Mix crispy coating ingredients. It should be prepared 15 minutes in advance before use.
4. Add salad sauce into mango and banana, and mix well. Wrap it in waffle paper, seal with egg white, and coat with crispy coating paste, and then deep fry in oil until golden brown.

Deep-fried Goose Meat 燒雁鵝 (p34)

☑ 20 mins   🍴 4-6

Main ingredient: Goose

Ingredients:
1 goose in spicy sauce, Some sweet and sour chili sauce or preserved tangerine oil, Some cornstarch

Egg mixture:
1pc egg, 3 tbsp cornstarch

Procedures:

1. Bone goose in spicy sauce.
2. Mix ingredients of egg mixture well. Coat goose meat and skin with egg mixture.
3. Coat with some cornstarch.
4. Heat some oil in wok. Put the whole piece of goose meat into it carefully, and deep fry until golden brown. Take it out and drain.
5. Chop the goose meat into pieces. Arrange them on plate and serve.

Fish Fillet with Nori 麵拖魚條 (p36)

Main ingredient: Fish; The minor: Nori

☑ 10 mins   🍴 4-6

Ingredients:
300g fish fillet, 20g nori (crispy seaweed)

Crispy coating batter:
½ cup deep-fried powder, 1 tbsp custard powder, 2 tsp glutinous rice powder, ½ cup cold water, 3 tbsp oil

Marinade Ingredients:
1 tsp chicken powder, 2 tsp lemon juice, 1 tbsp egg white, 1 tsp cornstarch, 1 tbsp cooked oil

Procedures:

1. Wash nori, pat dry, and then deep fry in hot oil for 20 seconds. Take it out and drain. Mash and set aside.
2. Cut fish fillet into thick strips. Add in marinade ingredients, and marinate for 15 minutes.
3. Mix crispy coating ingredients. Add in mashed nori, and mix well. It should be prepared 20 minutes in advance before use.
4. Deep fry in hot oil until 80% cooked.
5. Coat fish strips with crispy coating batter, and then deep fry until golden brown and firm.

Deep Fried Small Chicken 脆炸子雞 (p38)

Main ingredient: Small Chicken

☑ 40-45 mins   🍴 2-4

Ingredients:
1pc (1050g) chicken

Sauce ingredients for skin:
1 tbsp rose essence wine, 2 tbsp vinegar, 1 tbsp maltose, 1 tbsp cornstarch

White spicy sauce Ingredients:
4pcs aniseed, 1 tsp clove, 10g dahurian angelica, 2pcs licorice root, 1pc dried tangerine peel, 20g sand ginger, 2pcs herb, 1 tsp white pepper, 2pcs ginger, 3pcs dried shallot, 80g salt, 60g sugar, 1 tsp Aji-no-moto, 2-3ml water

Procedures:

1. Put white spicy sauce ingredients into pot, and add in water (the water level should be slightly higher than the height of chicken). Boil over medium heat for about 15 minutes.
2. Put chicken into boiled white spicy sauce. Turn the heat off and soak for about 15 minutes. Take it out, and let cool.
3. Boil sauce ingredients for skin (i.e. rose essence wine, vinegar and maltose). After cooled, stir in cornstarch. Spread evenly onto chicken for 2-3 times.
4. Hang chicken in windy place for 2-3 hours. Let dry.
5. Boil oil until 70% hot, and put skimmer / colander into it for a while. Place chicken into skimmer, and splash oil onto chicken (use "oil splashing method") until golden red and cooked.
6. When the skin is crispy, chop the chicken into pieces quickly and arrange on plate.

Roast Squab 紅燒乳鴿 (p40)

Ingredients:
2pcs squab

☑ 30 mins   🍴 4-6

White spicy sauce ingredients and sauce ingredients for skin:
The ingredients are the same as Deep Fried Small Chicken. Please refer to the previous recipe.

Procedures:

1. Wash squab, and remove its lung. Put into boiled white spicy sauce. Turn the heat off.
2. Soak for about 15 minutes. Take it out, and use hot water to wash away its oil.
3. Drain, and spread sauce ingredients for skin evenly onto squab. Hang it in windy place, and let dry.
4. Boil large amount of oil until 70% hot. Turn to slow heat, and then deep fry squab until the colour of squab changes. Turn to medium heat, and then deep fry squab until golden red. Serve with salt and worcestershire sauce.

Indonesian Chicken Rolls 印尼雞卷 (p42)

Main Ingredients: Mixed ingredients

Skin Ingredients:
2pcs egg, 2 tbsp sugar, 1 cup milk (about 250g), 80g flour, 30g oil

Cutlets Batter:
2pcs egg, 2 tbsp flour

Seasonings:
¼ tsp each of chicken powder and sugar, A little pepper, Bread crumb

Fillings:
½ cup each of diced chicken, diced mushroom and diced onion, ¼ cup mixed vegetables, 1pc bacon (diced), 1pc egg (cooked and diced)

White sauce Ingredients:
30g butter, ¼ cup flour, ½ cup water, 1 tbsp evaporated milk, ¼ tsp salt, 1 tsp chicken powder

☑ 30 mins ▨ 6-8

Procedures:
1. Heat some oil in wok. Shallow fry onion and bacon, then chicken, vegetables, mushroom, etc. Stir in seasonings, and add in diced egg at last. Cook well to make fillings. Set aside.
2. Shallow fry flour with butter. Stir in water slowly, and cook over slow heat until all water is added. Stir in evaporated milk and fillings. Let cool, and set aside.
3. Mix egg and sugar, then add in flour one by one, and mix well. Stir in oil, and shallow fry to make thin round / circle. Wrap with fillings to form chicken rolls. Set aside.
4. Coat wrapped chicken rolls with cutlets batter, and then bread crumb.
5. Deep fry in hot oil until golden brown. Serve.

Spicy Mutton Rolls 麻辣羊肉春卷 (p44)

Main ingredient: Mutton

Ingredients:
150g minced mutton, ¼ cup onion (chopped), 10pcs spring roll sheet

Egg batter (mixed well):
1 egg yolk, 1 tbsp cornstarch

Seasonings:
¼ tsp chili powder, ½ tsp caraway powder, ¼ tsp peppercorn powder, 1 tsp soy sauce, 1 tsp chicken powder, ½ tsp sugar, 1 tbsp cornstarch, 1 tsp sesame seed oil

☑ 10 mins ▨ 4-6

Procedures:
1. Place mutton and chopped onion into mixing bowl. Add in seasonings, and mix well.
2. Spread flat of spring roll sheet, and add in some fillings. Fold both ends towards the centre, and roll it up to form cylinder. Seal opening with little egg batter.
3. Deep fry in 80% hot oil until golden brown. Serve.

Crispy Liver Rolls 脆香鳳肝卷 (p46)

Main Ingredients: Mixed ingredients (i.e. the amounts of main and minor ingredients are similar)

Ingredients:
120g chicken liver or pig liver (marinated with ginger juice), 2pcs Chinese mushroom (soaked and shredded), 20g fine pickled vegetables (diluted and shredded), 120g celery (shredded), 40g leek sprout (shredded), 2pcs bean curd sheet (divided into 4), 1pc garlic (mashed), 1pc shallot (mashed)

Seasonings:
½ tsp chicken powder, 1 tbsp oyster sauce, 1 tsp sugar, ½ tsp five spice powder, A little sesame seed oil, A little pepper, 1 tsp cornstarch, 2 tbsp water

Crispy coating batter:
50g baking powder, ½ tsp custard powder, 2 tsp glutinous rice powder, ¼ tsp baking soda, ⅓ cup water, 1 tbsp oil

☑ 10 mins ▨ 4-6

Procedures:
1. Mix ingredients of crispy coating batter, set aside for 15 mins.
2. Boil chicken liver or pig liver until just cooked. Cut into strips.
3. Heat some oil in wok. Shallow fry shallot, onion puree, and then shredded Chinese mushroom, pickled vegetables and celery. Add in seasonings, and mix well. Let cool. Wrap bean curd sheets with fillings to make spring rolls.
4. Heat some oil in wok. Coat spring rolls with crispy coating paste, and deep fry until golden brown. Cut into pieces, and serve.

Squid Ring Cutlets 吉列鮮魷環 (p48)

Main ingredient: Squid

☑ 10 mins   🍴 4-6

Ingredients:
2pcs fresh squid (about 300g), Some original flavour corn crisps and lemon (garnish)

Marinade Ingredients:
3 tbsp evaporated milk, ½ tsp spice salt, 1 tsp chicken powder, Dash of pepper

Egg batter (mixed well):
2pcs egg, 2 tbsp flour

Dipping:
Thousand Island dressing

Procedures:
1. Remove skin from squids and wash. Cut sideways into rings. Mix with marinade ingredients, and set aside for 15 minutes.
2. Crush corn crisps. Dip squid rings into egg batter, and then coat with crushed crisps.
3. Deep fry squid rings in hot oil (about 220℃) until golden brown. Serve with Thousand Island dressing.

Matsutake and Shrimp Salad 吉列靈芝菇鮮蝦沙律 (p50)

Main Ingredients: Matsutake, shrimp; The minor: Liver sauce, onion

☑ 10 mins   🍴 4-6

Ingredients:
300g Matsutake, 12pcs shrimp (medium size), ½pc onion (diced), 40g liver sauce (any type), ½pc melon, Some waffle paper, Some bread crumb, Some salad dressing

Marinade shrimp Ingredients:
1 tsp egg white, ¼ tsp cornstarch, ⅛ tsp salt

Egg batter:
2pcs egg, 4 tbsp cornstarch

Seasonings:
¼ tsp salt, ½ tsp chicken powder

Procedures:
1. Wash Matsutake and boil in water for a while. Remove shell from shrimp and intestine, and cut sideways at shrimp's tail. Mix with marinade ingredients, and deep fry.
2. Carve melon (with skin) into 12 small rhombus cups. Set aside.
3. Heat little oil in wok. Shallow fry onion and Matsutake, then stir in seasonings. Drain excess oil.
4. Add ½ onion and Matsutake into liver sauce. Wrap waffle paper with these fillings to make spring rolls. Seal the opening with little egg batter.
5. Coat wrapped Matsutake rolls with egg batter, and then bread crumb. Deep fry in 80% hot oil until golden brown.
6. Dice flesh of melon, and add in remaining Matsutake and salad dressing. Put into melon rhombus cups, and then add in deep-fried shrimps. Serve.

Oil Soaked Chick with Lemongrass 油浸香茅仔雞 (p52)

Main ingredient: Chick

☑ 15-18 mins   🍴 4-6

Ingredients:
1pc chick (400g)

Marinade Ingredients:
2 tsp salty chicken powder, 1 tsp soy sauce, 1 tsp rose essence wine, 1 tbsp cooked oil, 1pc lemongrass (cut / shredded, slightly pressed), 1pc aniseed

Dipping sauce:
½pc lime (cut into pieces), 1pc chili (diced), 1 tsp garlic, 1 tbsp fish sauce, 2 tbsp sugar

Procedures:
1. Wash chick, and remove its lung. Spread marinade ingredients onto exterior and interior of chick. Put lemongrass and aniseed into chick, and marinate for 1 hour.
2. Take out lemongrass and aniseed, and use absorbent paper to absorb excess water.
3. Hang chick in windy place for about 1 hour. Let dry.
4. Heat a large amount of oil until 80% hot. Deep fry chick for 2 minutes, then turn the heat off. Soak for 15 minutes, and then take it out. Let cool, and chop into pieces. Serve.

Oil Soaked Tench with Vinegar 油浸香醋鯇魚尾 (p54)

Main ingredient: Tench

Ingredients:
1pc tench (480g), 20g ginger (shredded), 20g scallion (shredded)

Seasonings:
2 tbsp Zhenjiang vinegar, 2 tbsp soy sauce, 1½ tbsp brown sugar, ⅛ tsp salt, ¼ tsp Aji-no-moto, ½ tsp cornstarch

☑ 10 mins 🍴 4-6

Procedures:
1. Wash tench and pat dry.
2. Soak in moderately high heat oil (80% hot) for 2 minutes. Turn off heat. Soak for 8 minutes, and take it out. Drain and place on plate.
3. Mix seasonings and boil, and then spread onto tench.
4. Spread shredded ginger and scallion on top, and then hot oil. Serve.

Fish with White Sauce 白汁海上鮮 (p56)

Main ingredient: Grouper

☑ 10 mins 🍴 4-6

Ingredients:
1pc grouper or other types of fish (600g), 1pc egg white, Pinch of shredded Jinhua ham, Pinch of parsley

Seasonings:
100g evaporated milk, 100g chicken broth, 1 tsp chicken powder, 2 tsp cornflour, 1 tsp ginger liquor

Procedures:
1. Wash fish and pat dry.
2. Pour a large amount of oil into wok (enough to completely submerse the fish).
3. Heat oil to moderately high heat (80% hot). Soak fish in oil for 2 minutes. Turn off heat. Soak in oil for 7-8 minutes (done when fish eyes protruding). Take it out, and drain well.
4. Mix seasonings and boil. Turn off heat. Add in egg white, pinch of oil and shredded ham. Pour onto fish. Garnish with some parsley. Serve.

Belly Pork with Ginger and Passion Fruit Jam 子薑百香果醬肉 (p58)

Main Ingredients: Mixed ingredients (i.e. the amounts of main and minor ingredients are similar)

☑ 10 mins 🍴 4-6

Ingredients:
300g belly pork (chopped), 80g peach (cut into pieces), 1pc red chili (diced), 60g green chili / pepper (cut into pieces), 20g sour ginger, 60g cornstarch (for coating)

Marinade Ingredients:
1 tsp soy sauce, 1pc egg yolk, ¼ tsp salt, ½ tsp chicken powder, Little sesame oil, Pinch of pepper, 1 tsp cornstarch

Seasonings:
4 tbsp water, 1 tbsp white vinegar, 2 tbsp brown sugar, 1 tsp Lea & Perrins sauce, 1 tsp chicken powder, 2 tbsp passion fruit jam, 4pcs Fernandes Cherry Bouquet (Kersen), 1 tsp cornstarch

Procedures:
1. Mix 1 tsp baking soda and 1 tbsp sugar with belly pork, and marinate for about 1 hour. Wash and drain well.
2. Mix marinade ingredients well. Add onto belly pork, and marinate for over 15 minutes.
3. Put marinade belly pork into sieve, and drain sauce away. Coat with cornstarch, and then deep fry in moderately high heat oil (80% hot oil) until golden brown and crispy. Drain well.
4. Sauté minor ingredients, add in seasonings and boil for a while until cooked. Turn heat off. Add in deep-fried belly pork, and stir quickly. Serve.

Deep-fried Eel 奇香鱔球 (p60)

Main ingredient: Eel

☑ 10 mins 🍴 4-6

Ingredients:
900g eel, Some toasted sesame seeds, Some cornstarch

Marinade Ingredients:
2 tsp soy sauce, 2 tsp salty chicken powder, 1 tsp rose essence wine, 1 tbsp cooked oil, 1 tbsp cornstarch

Seasonings:
3 tbsp salad dressing, 1 tsp custard powder, 4 tbsp water, 1 tsp condensed milk, 1 tsp lemon juice

Procedures:
1. Wash and bone eel, and then wash again and slightly cut squid-like pattern.
2. Mix with marinade ingredients, and then coat with cornstarch.
3. Deep fry in moderately high heat oil (80% hot oil) until golden brown. Drain well.

4. Boil seasonings, add in eel, and stir well quickly. Place on plate, and sprinkle some toasted sesame seeds on top. Serve.

Deep-fried Spicy Fish Fillets 椒麻魚塊 (p62)

Main ingredient: grass scrap

☑ 30 mins 🍴 6-8

Ingredients:
1200g grass scrap belly, A few slices of sectioned ginger (marinade fish fillet), 2 sprigs of sectioned spring onion (marinade fish fillet), ½ tsp Sichuan peppercorn powder (for finishing), 1 tbsp chopped ginger (for finishing), 1 tbsp chopped garlic (for finishing), 1 tbsp sesame oil

Spicy sauce:
5 tbsp soy sauce, 40g cane sugar, ½ tsp salt, 1 cup water

Fish fillet marinade Ingredients:
2 tbsp soy sauce, ½ tsp salt, 1 tbsp wine with ginger juice, 1 tbsp cooked oil

Spicy sauce Ingredients:
2pcs star anise (aniseed), 2 tsp Sichuan peppercorn, 2pcs ginger (slices), 1 sprig of spring onion

Procedures:
1. Wash fish fillet and chop into large pieces. Add in ginger, spring onion and fish fillet marinade ingredients, and toss the mixture. Marinate for 30 minutes, pat dry and set aside.
2. Heat oil and sauté spicy sauce ingredients. Add in spicy sauce and boil. Then simmer over low heat until only ½ cup of sauce left. Sieve / Filter sauce and set aside.
3. Heat wok and add some oil. When oil is hot, add in fish fillets and then deep fry until golden brown. Drain off excess oil and set aside.
4. Poach fish fillets in spicy sauce until sauce thickens. Take fish fillets out and set aside.
5. Heat 1 tbsp sesame oil. Add in Sichuan peppercorn powder, ginger and garlic. Sauté until aroma is released.
6. Add in fish fillets. Stir fry and serve.

Eggplant Sambal 參巴茄子 (p64)

Main ingredient: Eggplant; The minor: Tomato, onion

☑ 15 mins 🍴 4-6

Ingredients:
500g eggplant, 100g tomato, 1pc onion, 1 stalk parsley, Some toasted white sesame seeds

Spices:
70g red chili pepper, 20g small chili, 100g shallot, 50g garlic

Seasonings:
2 tbsp tomato paste, 1 tbsp Lea & Perrin hot pepper sauce, ½ tsp sugar and salt, 1 tsp chicken powder, ¼ cup water, 1½ tsp HP Sauce

Procedures:
1. Slice eggplant and then deep fry in hot oil for about 1-2 minutes. Remove and strain oil.
2. Slice onion, chop up parsley, and slice tomato.
3. Heat 3 tbsp oil, shallow fry / sauté spices until aroma is released. Add in onion and shallow fry until cooked. Stir in seasonings and cook well.
4. Put eggplant back into wok and mix well. Add tomato and cook for 2-3 minutes. Then add parsley, mix well and place on plate. Sprinkle some sesame seeds on top.

Sweet and Sour Shrimp 醋熘明蝦球 (p66)

Main ingredient: Shrimp; The minor: White fungus, bean

☑ 10 mins 🍴 4-6

Ingredients:
300g shrimp (middle-sized), 80g white fungus (soaked), 40g bean, 2pc garlic (slices), 2pc shallot (cut edges), Some cornstarch

Marinade Ingredients:
1 tsp egg white, ¼ tsp salt, 1 tsp cornstarch

Seasonings:
1 tsp Zhenjiang vinegar, 1 tbsp oyster sauce, 1 tsp soy sauce, 2 tbsp tomato sauce / paste, 1 tsp chili oil, 10g Fernandes Cherry Bouquet (Kersen), 1 tsp chicken powder, 2 tsp cornstarch, ½ cup water

Procedures:
1. Wash shrimps, remove their shells and intestines, but retain tails. Slice garlic shoot.
2. Mix marinade ingredients with shrimps, and coat with cornstarch. Heat oil and set aside.
3. Heat 2 tbsp oil, sauté garlic slices, shallot and garlic shoot, and add in white fungus. Shallow fry for a while, and stir in seasonings. Cook well.
4. Put shrimps back into wok. Shallow fry for a while, mix well and place on plate.

Yellow Croaker with Pine Nut 松子黃魚 (p68)

Main ingredient: Yellow croaker; The minor: Pine nut, tomato, bean

☑ 15 mins 🍴 4-6

Ingredients:
1pc yellow croaker, 1 tbsp pine nut, ¼ cup diced tomato, 1 tbsp green pea / bean, Some cornstarch

Marinade Ingredients:
1 tsp lemon juice, 1 tbsp cooked oil, ½ tsp salt, ½pc egg

Seasonings:
½ cup tomato sauce / paste, 1½ tbsp sugar, 1 tsp custard powder, ½ tsp chicken powder, Little sesame oil, ¼ cup water

Procedures:
1. Wash and bone fish. Marinate for 15 minutes, and then coat with cornstarch. Set aside.
2. Deep fry fish over moderately high heat oil until golden brown. Arrange on plate.
3. Boil seasonings for a while. Add in diced tomato and green pea. Stir well. Rinse sauce onto fish. Sprinkle some pine nuts on top. Serve.

Stuffed Black Pomfret with Belachan 馬拉盞煎黑鯧 (p70)

Main ingredient: Black pomfret

☑ 20 mins 🍴 4-6

Ingredients:
1pc black pomfret (about 400g)

Seasonings:
1½ tsp sugar, 1 tsp lemon juice, 1 tsp Lea & Perrin hot pepper sauce, ½ tsp salt

Sauce Ingredients:
3pc red chili pepper, 20g belachan, 4pcs each of garlic and shallot

Procedures:
1. Clean fish, and make slits on fish body to form sandwich pattern.
2. Crush all sauce ingredients, add in seasonings, and mix well. Then stuff belachan mixture into fish.
3. Pan fry fish over low heat until both sides turn golden brown. Serve hot.

Shallow-fried Sun-dried Oysters 乾煎生曬蠔豉 (p72)

Main ingredient: Sun-dried oyster

☑ 15 mins 🍴 4-6

Ingredients:
12pcs sun-dried oyster, Some cornstarch

Marinade Ingredients:
1 tbsp ginger liquor, 2 tsp honey, 1 tsp lemon juice, 1 tsp soy sauce, 1pc red chili (diced)

Procedures:
1. Wash sun-dried oysters. Soak in water overnight. Drain and pat dry.
2. Marinate oysters for 30 minutes.
3. Lightly coat with cornstarch.
4. Heat some oil in wok. Shallow fry oysters over low heat until both sides turn golden brown.

Shallow-fried Cattle Fish Cake with Radish 香煎蘿蔔墨魚餅 (p74)

Main Ingredients: Radish, minced cattle fish; The minor: Celery, chili

☑ 15 mins 🍴 4-6

Ingredients:
1pc radish (300g), 240g minced cattle fish, 2 tbsp celery (diced), 1pc chili (diced), 1 cup of water (for cooking radish)

Seasonings:
½ tsp salt, 1 tsp chicken powder, Pinch of pepper, Little sesame oil, 1 tbsp glutinous rice powder

Procedures:
1. Shred radish. Boil it in 1 cup of water for 10 minutes. Drain and let cool. Set aside.
2. Put minced cattle fish in bowl. Add in shredded radish, diced celery, diced chili and seasonings, and mix until sticky.
3. Spread little oil onto hands. Divide the above ingredient into portions, and knead to make balls. Slightly press.
4. Heat pan, and add in a large amount of oil. Shallow fry cattle fish cake over low heat until both sides turn golden brown. Simultaneously, spread hot oil onto cattle fish cake surface. After cooked, chop into pieces and arrange on plate.

Fish Pancake with Coriander 鍋貼香菜蛋包魚 (p76)

Main ingredient: Fish; The minor: Egg, coriander

☑ 8-10 mins   📷 4-6

Ingredients:
5pcs egg, 1pc yellow croaker, 2pcs coriander (chopped), 1pc red chili (diced), 1 tsp ginger rice, Some cornstarch

Marinade Ingredients:
1 tsp egg white, ½ tsp salt, ½ tsp cornstarch, 1 tbsp cooked oil

Seasonings:
½ tsp chicken powder, ½ tsp salt

Procedures:
1. Whisk egg, add in seasonings, and mix well. Set aside.
2. Wash fish, bone and dice. Mix with marinade ingredients and blanch in hot oil. Let cool, and set aside. Coat head, tail and fish bone with some cornstarch. Deep fry until golden brown, and arrange on plate.
3. Add fish meat, chopped coriander, diced red chili, and ginger rice into whisked egg. Mix well.
4. Heat pan, and add in some oil. Shallow fry the above ingredients over medium heat to form hot pancake shape.
5. Let cool. Cut into pieces, and arrange on deep-fried fish bone. Serve.

Fried Tofu with Shrimp Roe 蝦子鍋塌豆腐 (p78)

Main ingredient: Bean curd; The minor: Shrimp roe

☑ 15 mins   📷 4-6

Ingredients:
1 packet hard tofu, 2pcs egg yolk, Some cornstarch, Some shrimp roe

Marinade Ingredients:
1 tsp chicken powder

Seasonings:
1 tsp chicken powder, ⅓ cup of stock

Procedures:
1. Pat dry tofu with cloth, and then cut into 6 pieces. Whisk egg yolk.
2. Coat tofu with cornstarch and then egg liquid.
3. Put it in hot oil, and pan-fry over medium heat until both sides turn golden brown.
4. Mix seasonings, pour it into pan on the side and cook over low to medium heat until sauce is reduced (turn over tofu once in between).
5. Transfer to plate, and sprinkle some fried shrimp roes on top. Serve.

Stir-fried Grouper Fillets with Mushrooms and Assorted Peppers
香菇彩椒炒斑柳 (p80)

Ingredients:
240g grouper fillet, 3pcs shiitake mushrooms, ½pc red pepper, yellow pepper and green pepper, 1pc garlic (sliced), 1 tsp Shaoxing wine

☑ 10 mins   📷 4-6

Seasonings:
1 tsp chicken powder, A little sesame oil, A pinch of pepper, 4 tbsp water, 1 tsp cornstarch

Marinade Ingredients:
1 tsp egg white, ½ tsp cornstarch, ½ tsp chicken powder, 1 tbsp cooked oil

Procedures:
1. Cut grouper fillet into strips, mix with marinade ingredients, then blanch in hot oil and set aside.
2. Soak mushrooms until softened, and then cut into strips. Cut red, yellow and green pepper into strips.
3. Stir-fry garlic slices in hot wok. Add in mushrooms and stir-fry for a while, and then add in red, yellow and green pepper. Stir-fry thoroughly; and then splash in Shaoxing wine.
4. Add in seasonings and grouper fillets, and stir-fry well. Transfer to plate and serve.

Spicy Stir-fried Chicken 麻辣雞丁 (p82)

Main ingredient: Chicken; The minor: Peanut, French bean, dried chili

☑ 25 mins   📷 6-8

Ingredients:
60g hot and spicy peanut, 80g garlic shoot or French bean (diced), 300g chicken thigh meat, 10g dried chili, 2 tsp minced garlic, 1 tsp minced ginger, Some cornstarch

Chicken marinade Ingredients:
½pc egg yolk, 2 tsp soy sauce, 1 tsp oyster sauce, ½ tbsp cornstarch, 1 tbsp cooked oil

Seasonings:
½ tbsp soy sauce, ½ tsp Zhenjiang vinegar, ½ tsp brown sugar, ½ tsp cornstarch, 2 tbsp water

Procedures:
1. Dice chicken meat, mix with marinade ingredients, and then coat with cornstarch before deep-frying.
2. Heat some oil to moderately high heat (80% hot), and then deep fry chicken until golden brown. Drain well.
3. Heat little oil, add in chili, minced garlic and ginger, and stir-fry. Then add in chicken, spicy peanut and garlic shoot / bean.
4. Add in seasonings, and stir-fry. Place on plate, and serve.

Stir-fried Eel Shreds 炒鱔糊 (p84)

Main ingredient: Eel; The minor: Chinese yellow chive

☑ 10 mins 🍴 4-6

Ingredients:
480g yellow eel, 2 tsp minced ginger, 2 tsp Shaoxing wine, 80g Chinese yellow chive, 3pcs garlic (finely chopped)

Marinade Ingredients:
1 tsp soy sauce, ¼ tsp salt, Some pepper powder, Some sesame oil, 1 tsp cornstarch, 1 tbsp cooked oil

Seasonings:
3 tbsp soy sauce, ½ tsp chicken powder, 1½ tsp cornstarch, 1¼ tsp sugar, Some sesame oil, ⅓ cup of water

Decoration:
Some pepper

Procedures:
1. Bone yellow eel. Use hot water to wash sticky liquid of eel. Cut into strips of 1½ inch long.
2. Mix with marinade ingredients. Heat oil to moderately high heat (80% hot), and set aside.
3. Heat 2 tbsp oil, sauté minced ginger, and splash in wine. Put eel back into wok, and shallow fry over high heat.
4. Add in yellow chive, and then seasonings. Stir-fry well.
5. Make a small hole in the middle, and add in chopped garlic.
6. Heat some sesame oil and 2 tbsp oil in wok, and spread onto chopped garlic. Sprinkle pepper powder on top. Mix well when serve.

Stir-fried Frog in Taro Bird Nest 玉種藍田 (p86)

Main Ingredients: Frog, Chinese kale / broccoli; The minor: Salted fish

☑ 8-10 mins 🍴 4-6

Ingredients:
400g frog, 240g Chinese kale / broccoli, 40g salted fish, 4pcs ginger (slices), 6pcs carrot (slices), 1 tbsp ginger juice, 1pc taro bird nest / basket

Marinade Ingredients:
1 tsp oyster sauce, 2 tsp soy sauce, ½ tsp chicken powder, Some sesame oil, Some pepper powder, 1 tbsp cornstarch

Seasonings:
2 tsp oyster sauce, 1 tsp soy sauce, 1 tsp chicken powder, Some sesame oil, Some pepper powder, 1 tsp cornstarch, 4 tbsp water

Procedures:
1. Wash and chop frog into pieces. Marinate for 15 minutes. Parboil in oil for use.
2. Cut Chinese kale into strips of 1.5 inch long. Blanch, rinse in cold water, and set aside.
3. Steam salted fish until done. Dice finely.
4. Heat some oil in wok. Sauté ginger, then add in frog, Chinese kale and salted fish. Stir-fry thoroughly. Add in seasonings, and mix well. Serve in taro bird nest.

Bean Sprout with Minced Pork 豆芽鬆 (p88)

Main ingredient: Bean sprout; The minor: Pork

☑ 15 mins 🍴 4-6

Ingredients:
640g bean sprout, 240g minced pork, 1pc celery (diced), 1 tsp garlic (finely chopped)

Marinade Ingredients:
1 tsp each of soy sauce, oyster sauce and oil, ½ tsp cornstarch, ¼ tsp sugar, Some pepper powder

Thickening sauce:
1 tbsp oyster sauce, 1 tsp each of soy sauce and cornstarch, ½ tsp each of sugar and chicken powder, Some sesame oil and pepper powder, 4 tbsp water

Procedures:
1. Mix marinade ingredients with pork. Marinate for a while, and set aside.
2. Remove the end of bean sprout, and then divide stalk into two. Mash / Chop head of bean sprout.
3. Heat head of bean sprout until dry. Add in 1 tbsp oil, and stir-fry. Then add in stalk and stir-fry until 70% cooked. Set aside.

4. Heat wok with some oil, and sauté chopped garlic. Add in pork and stir-fry thoroughly. Then add in bean sprout, diced celery and thickening sauce. Stir well and serve.

Salted Fish with Mixed Vegetables 馬友鹹魚炒雜菜 (p90)

Main Ingredients: Mixed ingredients (i.e. the amounts of main and minor ingredients are similar)

☑ 8-10 mins 📖 4-6

Ingredients:
40g salted fish, 160g asparagus, 80g bean sprout, 80g egg plant, 80g bamboo shoot, 80g honey bean, 80g yellow pepper, 1 tsp garlic (shredded), 1 tsp ginger (shredded)

Seasonings:
1 tsp chicken powder, 2 tsp cornstarch, 2 tbsp water

Procedures:
1. Dice salted fish, and mix with marinade ingredients.
2. Wash all vegetables and chop well. Parboil egg plant for a while, and set aside.
3. Heat 2 tbsp oil, sauté garlic and ginger. Then add in salted fish and shallow fry until cooked.
4. Add in all vegetables and shallow fry over high heat. Then add in bean sprout and fried egg plant, and shallow fry for about 20 seconds.
5. Add in seasonings, and stir well. Serve.

Sautéed Shredded Meat with Bean Sprout 金銀芽炒滑肉絲 (p92)

Main Ingredients: Bean sprout, sliced meat; The minor: Leek, bamboo shoot

☑ 8-10 mins 📖 4-6

Ingredients:
160g steak, 60g bamboo shoot, 300g bean sprout, 20g leek, 1pc garlic (slice), 2 tbsp carrot (shredded)

Marinade ingredients A:
½ tsp soda powder, 2 tbsp water

Marinade ingredients B:
1 tsp egg white, ¼ tsp MSG, ½ tsp cornstarch, 2 tsp water, 1 tbsp cooked oil

Seasonings:
1 tsp chicken powder, ¼ tsp salt, A pinch of pepper, 4 tbsp water, ½ tsp sugar, Little sesame oil, 1 tsp cornstarch

Procedures:
1. Slice steak into strips. Mix with marinade ingredients A for 20 minutes. Wash and drain well. Mix with marinade ingredients B. Parboil for a while, and set aside.
2. Slice bamboo shoot into strips. Blanch, wash with cold water and drain well. Chop leek.
3. Heat wok with 2 tbsp oil. Sauté sliced garlic. Shallow fry shredded bamboo shoot for a while, and then add in bean sprout until cooked. Add in shredded meat, leek and seasonings, and stir well. Serve.

Egg White with Clam 鮮蛤蜊炒牛奶 (p94)

Main Ingredients: Egg white, milk; The minor: Clam, pine nut

☑ 10-15 mins 📖 4-6

Ingredients:
200g milk, 80g clam, 20g Jinhua ham, 150g egg white, 25g pine nut, 2 tbsp lard (add at two different times, stir fry egg white), 2 tbsp oil (add at two different times, stir fry egg white)

Milk whey:
50g milk, 20g cornstarch

Seasonings:
¼ tsp Aji-no-moto

Procedures:
1. Stir well milk whey. Wash and blanch clam, and drain well. Dice Jinhua ham, and deep fry pine nut (until crispy).
2. Put egg white in mixing bowl. Add in seasonings, and mix well.
3. Heat egg white from medium heat to high heat. However, you must not over-cook it. Take it away from heat, and let cool. Add in milk whey, clam, Jinhua ham, etc. and mix well to form milk ingredients.
4. Heat wok with 1 tbsp oil and 1 tbsp lard over medium heat (until moderately high heat / 60% hot). Add in milk ingredients, and stir fry from one direction. Turn the wok upward when stir frying.
5. Add in 1 tbsp oil. When milk ingredients become semi-coagulated, add in pine nut, and 1 tbsp lard. Stir fry until coagulated. Arrange on plate to form hill shape.

Scrambled Egg with Beef 滑蛋炒牛肉 (p96)

Main Ingredients: Mixed ingredients (i.e. the amounts of main and minor ingredients are similar)
Ingredients:
120g beef, 5pcs egg, 2 tbsp spring onion

Marinade ingredients A:
½ tsp baking soda, ¼ cup of water

Seasonings:
½ tsp chicken powder, 1 tbsp ginger juice, ½ tsp salt, 1 tbsp white wine

Marinade ingredients B:
1 tsp soy sauce, ¼ tsp sugar, ½ tsp chicken powder, ½ tsp cornstarch, ½ egg yolk, 1 tsp water, 1 tbsp cooked oil

☑ 8-10 mins   🍴 4-6

Procedures:
1. Slice beef, and mix with marinade ingredients A for 1 hour. Wash, and drain well. Mix with marinade ingredients B. Set aside.
2. Whisk egg, add in seasonings, and mix with diced spring onion.
3. Parboil beef. Drain well. Add into whisked egg, and mix well.
4. Heat wok with 4 tbsp oil, and add in all other ingredients. Stir fry over high heat until semi-coagulated. Serve.

Pork Roll 醬爆春花卷 (p98)

Main ingredient: Pork; The minor: Bamboo shoot, chestnut, walnut

☑ 10-15 mins   🍴 4-6

Ingredients:
160g pork, 60g bamboo shoot (sliced), 4pcs chestnut, 80g sweet bean, 80g mushroom (cut into half), 6pcs carrot (sliced), 4pcs ginger (sliced), 1pc red chili (chopped)

Seasonings:
1 tsp soy sauce, 1 tsp oyster sauce, ½ tsp sugar, 1 tsp chicken powder, 1 tsp cornstarch, Little sesame oil, A pinch of pepper, 5 tbsp water

Fillings:
40g walnut (chopped), 1 tbsp roasted sesame

Sauce:
2 tsp chili bean sauce, 1 tsp garlic (chopped)

Marinade ingredients A:
½ tsp soda powder, 2 tsp sugar, 4 tbsp water

Marinade ingredients B:
1 tbsp egg white, ¼ tsp salt, ¼ tsp sugar, 1½ tbsp cornstarch

Procedures:
1. Chop pork into 12 pieces. Marinate with marinade ingredients A for 40 minutes. Wash and drain well.
2. Marinate pork with marinate ingredients B for 15 minutes. Set aside.
3. Dust some cornstarch on plate, place pork on it, add in fillings, and wrap to form roll.
4. Blanch bamboo shoot, sweet bean and mushroom. Wash with cold water, and drain well.
5. Coat meat roll with cornstarch, and deep fry in moderately high heat oil (80% hot) until golden brown and crispy. Take it out, and drain well.
6. Heat wok with some oil, sauté sliced ginger and sauce. Add in all other ingredients (except meat roll), sprinkle wine, and stir in seasonings. Cook for a while, and add in meat roll. Stir well quickly, and arrange on plate.

Spicy Frog 麻辣芥醬田雞 (p100)

Main ingredient: Frog; The minor: Chinese broccoli

☑ 10-15 mins   🍴 4-6

Ingredients:
2pcs frog, 3pcs Chinese mushroom, 160g Chinese broccoli, 8pcs carrot (sliced), 2pcs garlic (sliced), 4pcs ginger (sliced), 1 tbsp ginger wine & ½ tsp sugar (mix well)

Marinade Ingredients:
1 tsp ginger wine, 1 tsp soy sauce, 1 tsp oyster sauce, ½ tsp sugar, 1 tbsp cooked oil, 2 tsp cornstarch

Seasonings:
1 tsp sesame sauce, 1 tsp mustard, A pinch of pepper, ½ tsp sugar, 4 tbsp water, 1 tsp chili oil, Little sesame oil, 1 tsp chicken powder, 1 tsp cornstarch

Procedures:
1. Wash frog and chop into pieces. Marinate for 20 minutes. Parboil until 80% cooked.
2. Soak Chinese mushroom and slice. Chop Chinese broccoli.
3. Heat wok with 2tbps oil, sauté sliced ginger, sprinkle ginger wine, add in Chinese broccoli and mushroom. Stir fry for a while, and add in 2 tbsp water. Take it out and drain well.
4. Heat wok with some oil, sauté sliced garlic. Add in frog, and stir fry for a while over high heat. Stir in seasonings and Chinese broccoli. Arrange on plate, and serve.

Deep-fried Golden Ball 油爆金球 (p102)

Main Ingredients: Mixed ingredients (i.e. the amounts of main and minor ingredients are similar)

☑ 10-15 mins 🍴 4-6

Ingredients:
6pcs chicken wing, 3pcs duck kidney, 6pcs shrimp (medium size), 80g celery, 80g white bamboo shoot, 6pcs carrot (sliced), 2pcs garlic (sliced)

Marinade Ingredients:
1 tsp soy sauce, 1 tsp oyster sauce, ½ tsp chicken powder, 1 tsp cornstarch, 1 tbsp cooked oil, 1 tsp water, 1 tsp rose essence wine

Marinade ingredients for shrimp:
1 tsp egg white, ⅛ tsp salt, ½ tsp cornstarch, 1 tbsp cooked oil

Seasonings:
1 tsp oyster sauce, 1 tsp soy sauce, 1 tsp chicken powder, Little sesame oil, A pinch of pepper, 1 tsp cornstarch, 4 tbsp water

Procedures:
1. Bone chicken wing and turn upside down. Slightly slice duck kidney (with pattern), marinate and parboil. Set aside.
2. Remove shell from shrimp and cut into half. Mix with marinade ingredients for shrimp, parboil and set aside.
3. Chop celery into pieces, and cut the edge of white bamboo shoot. Blanch, wash with cold water, and drain well.
4. Heat 2 tbsp oil, sauté sliced garlic, add in celery, white bamboo shoot and sliced carrot. Stir fry well.
5. Put all ingredients back to wok. Stir fry over high heat, add in seasonings, and mix well. Arrange on plate, and serve.

Stir-fry Duck Meat with Herbs 爆炒香草鮮鴨片 (p104)

Main ingredient: Duck chest; The minor: Celery, pepper, ginger

☑ 10-15 mins 🍴 4-6

Ingredients:
1pc dust chest (160g), 80g celery, 40g red pepper, 40g yellow pepper, 2pcs sweet basil (get leaf), 1pc garlic (sliced)

Marinade Ingredients:
2 tsp soy sauce, 1 lemongrass (slightly press), 1 tsp honey, ½ tsp constarch, 1 tsp cooked oil

Seasonings:
1 tsp soy sauce, ½ tsp sugar, Little oil, ⅛ tsp thyme powder, 1 tsp oyster sauce, 1 tsp chicken powder, 1 tsp cornstarch

Procedures:
1. Wash duck chest and marinate for 30 minutes. Parboil until golden brown. Take it out, let cool, and slice into thick pieces.
2. Cut celery, red and yellow pepper into strips.
3. Heat 2 tbsp oil, sauté sliced garlic, add in celery, red pepper, yellow pepper, duck chest, sweet basil, etc. Stir fry for a while over high heat. Stir in seasonings, and mix well. Arrange on plate, and serve.

Spicy Chicken with Garlic 麻辣蒜香雞 (p106)

Main ingredient: Chicken; The minor: Chili, garlic shoot

☑ 10-15 mins 🍴 4-6

Ingredients:
200g chicken leg meat, 160g garlic shoot, 40g red pepper, 3pcs dried chili, 6pcs garlic, 1 tsp ginger puree, Some cornstarch

Marinade Ingredients:
½pc egg yolk, 1 tbsp soy sauce, ½ tsp parsley powder, ¼ tsp salt, ¼ tsp sugar, 2 tsp cornstarch, 1 tsp cooked oil

Seasonings:
1 tsp chicken powder, 2 tsp Zhenjiang vinegar, ¼ tsp sesame oil, 1 tsp cornstarch, 2 tsp brown sugar, 1 tbsp soy sauce, ½ tsp chili oil, 4 tsbp water

Procedures:
1. Chop garlic shoot, red pepper, and dried chili
2. Wash chicken meat, cut into pieces, mix with marinade ingredients, and coat with some cornstarch before deep frying.
3. Heat oil to moderately high heat (80% hot), deep fry chick meat until golden brown. Take out and drain excess oil.
4. Heat 2 tbsp oil, sauté garlic, ginger puree and dried chili, add in red pepper and garlic shoot. Stir fry for 20 seconds.
5. Put chicken meat back to wok, stir fry over high heat, and stir in seasonings. Stir fry until little sauce left. Arrange on plate, and serve.

Crispy Purple Sweet Potato 拔絲紫心薯 (p108)

Main ingredient: Purple sweet potato

☑ 10 mins ⬛ 4-6

Ingredients:
200g purple sweet potato, 100g white sugar, 36g water

Crispy coating paste:
160g deep-fried powder, 1 tbsp glutinous powder, 10g custard powder, Some water, 30g oil

Procedures:
1. Peel purple sweet potato skin, and dice.
2. Put crispy coating ingredients into mixing bowl, and add in some water to form paste. Add in oil at last, and mix well. Set aside.
3. Coat purple sweet potato with crispy coating paste, and deep fry in moderately high heat oil (80% hot) until golden brown. Take it out, and drain well.
4. Put white sugar and water into saucepan. Boil over low to medium heat, and stir well by ladle.
5. When sugar water (syrup) changes from light yellow with large bubble to dark yellow with small bubble, use chopstick to drop syrup into iced water. When it becomes semi-coagulated, remove from heat (fire). Add in deep-fried purple sweet potato, and mix well quickly.
6. Grease plate (with oil), and place main ingredients coated with syrup onto it. Use chopstick to get main ingredients and sugar. Soak in iced water for a while to form crispy dessert.

Sweet Taro 反沙芋 (p110)

Main ingredient: Taro

☑ 10-15 mins ⬛ 4-6

Ingredients:
200g taro, 6pcs shallot, 90g sugar, 30g water, Some oil (cover all other ingredients)

Procedures:
1. Cut shallot into half, and deep fry until golden brown and crispy. Take it out, and drain well.
2. Peel taro skin, cut into strips, and deep fry (in oil which is used for deep frying shallot) until golden brown and firm. Take them out, and drain well.
3. Put white sugar and water into saucepan. Boil over low to medium heat until bubble forms. When sugar crystals agglomerate and become large enough, add in deep-fried taro strips, and stir well quickly. Turn heat off.
4. After cool, taro strips will spread out. Arrange on plate, and serve.

Deep-Fried Egg White Puff with Red Bean and Banana 高麗豆沙香蕉 (p112)

Main ingredient: Egg white; The minor: Red bean paste

☑ 10 mins ⬛ 8

Ingredients:
5pcs egg white (about 150g), 20g flour, 40g cornflour (sifted), 40g red bean paste, ½pc banana (diced)

Procedures:
1. Divide red bean paste into 8 portions. Add a dice of banana into each portion of paste, and knead it to form ball shape.
2. Whisk egg white into thick cream. Add in sifted flour and cornflour, and blend into egg white batter.
3. Use ice-cream ladle to take half scoop of egg white batter, and then add a tiny banana red bean ball into it.
4. Fill ladle with egg white batter. Place it into pan with warm oil and deep fry over low heat until cooked well and in light yellow colour. Take out fried puff and drain out excess oil. Dust some icing sugar over it, and serve hot.